Audio Programming for Interactive Games

This book is dedicated to my family, from whom I gain my greatest inspiration, strength and comfort. I am truly blessed.

Audio Programming
for Interactive Games

Martin D. Wilde

AMSTERDAM • BOSTON • HEIDELBERG • LONDON • NEW YORK • OXFORD
PARIS • SAN DIEGO • SAN FRANCISCO • SINGAPORE • SYDNEY • TOKYO
Focal Press is an imprint of Elsevier

ELSEVIER

Focal Press

Focal Press
An imprint of Elsevier
Linacre House, Jordan Hill, Oxford OX2 8DP
200 Wheelers Road, Burlington, MA 01803

First published 2004

British Library Cataloguing in Publication Data
A catalogue record for this book is available from the British Library

Library of Congress Cataloguing in Publication Data
A catalogue record for this book is available from the Library of Congress

ISBN 0 240 51941 8

For information on all Focal Press publications visit our
website at www.focalpress.com

Typeset by Newgen Imaging Systems (P) Ltd., Chennai, India
Printed and bound in Great Britain by Biddles Ltd, King's Lynn, Norfolk

Contents

Preface

Over the past several years, the basic technical challenges of delivering audio under computer control have been progressively met in contemporary entertainment devices. From desktop PCs and game consoles, and increasingly to all manner of handheld, wireless and mobile devices, highly sophisticated entertainment appliances that can play back sound with great quality, capacity and proficiency surround us. Their wide distribution is also due to the popularity of computer games and entertainment software.

Computer games are inherently interactive experiences, with the player's choices and skill being a major determinant in the game's final outcome. The same cannot generally be said of a game's soundtrack, which is often experienced linearly; you listen to it beginning to end, akin to one's enjoyment of a conventional audio CD. In addition, the sound is usually not controlled by the artists who created it. This power rests rather in the hands of the game programmer who may or may not know anything about audio, let alone the audition of the composer.

This book picks up the story of game audio in a period of significant transition. Freed by and large from the low-level concerns of computer audio playback, we are only scratching the surface of the creative potential these audio systems present. Computer Music grew from the desire to have technology serve musical purposes. In games, it is the other way around. To move game audio technology forward, the power of game audio systems must rest in the hands, ears, hearts and minds of the musicians and audio artists who use them. We now have the opportunity, means and obligation to creatively and musically fashion immersive, interactive, nonlinear auditory experiences in a game context. This book explores the musical controls needed to do so, and details the implementation of an audio software system that allows an audio artist to maintain direct control over the composition and presentation of a clean and seamless interactive soundtrack in a game setting.

This book is also concerned with building bridges between the artistic and technical communities, and necessarily crosses a lot of boundaries. It introduces game audio to many different target populations at all levels of experience, responsibility, interest and authority. For game developers and audio programmers, it gives the technical details of an interactive audio system. For game audio composers and sound designers, it describes an interactive audio system they can control and utilize to their own musical purposes. No longer must they rely on a programmer to realize their audition for the project. For game producers and publishers, it breaks interactive audio down into plain English, and makes the business case for cross-platform audio engines. Schools offering classes or degree programs in game development will find in this book an introduction to the use and support of audio in games. And finally, all students of the audio arts looking to learn, appreciate and extend their craft into multimedia and games will appreciate the information on the audio technology and interactive music concepts contained herein.

Whatever your perspective, I hope you all enjoy it, and perchance learn a thing or two along the way.

May you create the fullest expression of yourself and your song!

Acknowledgments

No project of this magnitude is accomplished alone, and there are many who helped to make it possible. First and foremost, I would like to thank my family for putting up with my absenteeism over these past many months. Colette, Dexter, George and Elinora, I love you more than words can tell. Thank you, thank you, thank you for your support, encouragement and love as I plowed my way through this project. I am looking forward to my re-entry.

I would also like to thank those who contributed their time and energy to this book from a technological point of view. The Soundtrack Manager described in these pages has taken years to develop and bring together, but I am extremely proud of the final result. So to Steve, Steve, David, Todd, Fred, Rob and Mike, thank you for your comments, suggestions and expertise. I couldn't have done it without you.

To Doug who vocalized what I had been thinking about doing for quite awhile, and to Francis who got the wheels rolling, thank you. Thanks also to my editor, Beth, who demonstrated great faith, patience and trust. As deadlines loomed and passed, you were always positive and focused on making it happen.

And finally, thanks to Focal Press for giving me the opportunity to bring this knowledge and experience to light and to the world. With this book, it is my sincere wish that the flame of interactive music burn a little brighter, and a little stronger. May all who read this book find something within its pages that makes their lives a little bit easier or a little more interesting.

Introduction

Welcome! Thank you for reading this book. May it give you a perspective on game audio you might otherwise have missed in your travels through the weird and wonderful universe that is games.

This book takes a step-by-step approach to the development and implementation of an interactive game audio software system. The term "interactive audio" itself can be confusing, but in the context of computer games refers to sound that is controlled or modulated by the user in real-time as the game progresses. In the context of a game, therefore, we have to create a soundtrack on the fly based on the player's movements and actions. This is a tall order, because we have no way of knowing a priori how the game is going to unfold. But all of the auditory manipulations of a game must happen in a way that actively, musically and logically changes in response to the tension of the situation. This is the goal of the software Soundtrack Manager presented in these pages.

Each chapter in this book builds on the last. So at least on the first time through, I encourage you to read the chapters in the order presented. I have tried to structure this book so as to give an overview of the subject matter at the beginning of each chapter, diving into the coding details as necessary as the chapter progresses. If you are a programmer, you will benefit from following along further as I drill down into the depths of data structures and operations. If you are not a programmer, feel free to skip those sections. To all readers of this text, however, it is important to keep in mind, that the role of the machine is to serve a larger human and musical purpose. My perspective is that the expression of the code must support the expression of the art. Therefore, I endeavor throughout to discuss first the musical purpose and artistic need before going on to explain the technology.

The book itself

The first three chapters begin with a look at the historic second-class status of audio in games. We take a brief tour of some of the commercial audio APIs out there, and then turn our attention to a description of and introduction to the promising realm of interactive audio. In Chapters 4 and 5, we delve into the major audio resources of games, digital audio and MIDI. We begin to build an interactive Soundtrack Manager

Audio Programming for Interactive Games

with the goal of providing high-level music and audio services across platforms. Platform-independent code samples are first presented here to support, explain and demonstrate the interactive nature of the sound library.

The Soundtrack Manager really takes shape in the middle chapters of the book. Chapter 6 presents an entirely new and extended MIDI file format that allows for arbitrary start points, looping and branching. The two major audio resources of digital audio and MIDI are merged in Chapter 7 with the detailed implementation of a custom software wavetable synthesizer. Chapter 8 continues that story with the presentation of an optimized, low-level assembly language version of the wavetable mixer, with some additional signal processing effects thrown in for good measure. Chapter 9 shows how these powerful audio processes can be dynamically controlled in real-time for musical and dramatic responsiveness.

A second major goal of the Soundtrack Manager is to put all the power and functionality of the audio system in the hands of the audio artist. Chapter 10 shows how a compelling soundtrack that adapts and changes in response to the player's actions in real-time is constructed using a simple, yet powerful scripting language. Chapter 11 presents a way to coordinate the visual and auditory displays by associating many different kinds of audio content with game objects. Finally in Chapter 12, we take a look at where we are, and speculate on the work that still needs to be done.

Code whys and wherefores

All code for the Soundtrack Manager, its associated libraries and utilities is included on the companion CD. From a programming perspective, it is important to follow along in the code when reading some parts of this book. But this is not required, for several reasons. The first is that I personally find it tedious to read through pages and pages of code, especially if I'm reading the book for the first time. Just give me the overall picture. If I want to get into the code, tell me where it is, and I'll get to it when I have the time. Second, I do not present the code in line-by-line detail so much as I discuss the higher-level behavior of the operations being performed. I do this in parallel with the code. Those programmatically inclined individuals will definitely understand it much more by following along. But programming rigidity does not have to be reflected in the prose, and I hope not to lose too many of you in those sections. Should you find yourself nodding off or not understanding something, go on to the next chapter or section. I won't mind, and neither should you.

"The devil is in the details," as the saying goes, and the Soundtrack Manager couldn't do its job if it ignored them. It is no secret that programming requires great specificity. Machines are not at all forgiving of human vagaries, and they have to be given very deliberate and unequivocal instructions. This is the purpose of the CD. All the code necessary to configure and build the Soundtrack Manager is included there. All code is written in C, with some nascent ideas of objects thrown in.

The ultimate aim of this book is to balance the head and heart perspectives of an interactive audio system for multimedia applications. Technology is a seductive mistress that makes it easy for us to believe we can solve all our problems simply via the application of more technology. But without the guidance of the heart, projects can become confused and unfocused. I truly believe that a project can only succeed through the thoughtful application of technology. There is an appreciation and an implicit understanding that each part of us must be used and work together to create and participate in our success. It's a new (and yet paradoxically old) way of approaching the problems we face, and recognizes, honors and respects one's whole self.

I encourage you to keep that perspective in mind as you read through the book and examine the code. Use it as you like, manipulate and change things as you need, throw away the bits that don't suit your requirements, mold it to your own purposes. But above all, have fun! Create interesting games and captivating works of art. I'll be listening.

Good luck and good sound!

Martin D. Wilde
Chicago, Illinois
Fall 2003

1 Regarding audio in games – The curtain rises

Music was forced first to select artistically, and then to shape for itself, the material on which it works. Painting and sculpture find the fundamental character of their materials, form and colour, in nature itself, which they strive to imitate. Poetry finds its material ready formed in the words of language. Architecture has, indeed, also to create its own forms; but they are partly forced upon it by technical and not by purely artistic considerations. Music alone finds an infinitely rich but totally shapeless plastic material in the tones of the human voice and artificial musical instruments, which must be shaped on purely artistic principles, unfettered by any reference to utility as in architecture, or to imitation of nature as in the fine arts, or to the existing symbolic meaning of sounds as in poetry. There is a greater and more absolute freedom in the use of the material for music than for any other of the arts. But certainly it is more difficult to make a proper use of absolute freedom, than to advance where external irremovable landmarks limit the width of the path which the artist has to traverse. Hence also the cultivation of the tonal material of music has, as we have seen, proceeded much more slowly than the development of the other arts."

Helmholtz, "On The Sensations of Tone."

Video rules, dude

In the game of games, video is king. Video came first, and reigns unyieldingly. The very word "video" is first in the genre. It can be likened to an arrogant aristocrat; one that consumes a vast majority of resources while allowing it's lowly subjects only the most meager drops of time and attention they don't deserve. One of those subjects is audio. For years, it has barely been given 10% of the total memory, storage, processing, personnel, spending and marketing budget of a game. When asked to relinquish a few more resources or cycles for audio use, the response from the video team has always had the same dramatic tone: "Who are those who dare to interrupt our glorious rule with their ceaseless clamoring? Have they not the brightest star in the heavens to shine upon them? Is not the magnificent radiance of the light of 10 000 polygons more important than their wretched, guttural murmurings? What greater aspiration can there be but to gaze upon our wonderfulness and to bask in the glow of our visual renderings? Ye should be happy to eat the scraps that drop from our table, and to curl up like a dog at the feet of your master at the end of the day.

"You think I jest! Nay, knave. Do not speak such blasphemy, lest a mighty bolt of lightning strike you down, reducing thee to but the smallest, insignificant pile of ashes! Take that!"

To which comes the reply, "I'm sorry. Were you talking to me? I *see* a lightning storm, but I don't *hear* anything. And I always thought it was a *thunder*bolt. You know, the kind of sound that rumbles your guts, catapults you out of bed in the middle of the night, heart pounding in your throat." Or when you're out camping, lying alone in your tent in the middle of the night, you prick up your ears and listen to the howling of the wind, count the seconds between the flash and the thunder, and wait out the storm.

"But wait! Shh! What was that? . . . There it is again." A harrowing and mournful howl pierces the night air, and the tent walls vibrate with the sound. It has been ten, long years since last I heard that cry, and a chill runs up my spine. [Cue childhood theme.] I recalled my father's words, [under . . . sage, wizened platitude] and it was then that I knew my destiny. Ancient and primordial, the beast was coming for me. [Cue title music and swell!]

Wow! Wasn't that much better with the audio track than without it? And it was all of your own creation. Where did you go? Was it a haunted forest? A wind-swept precipice? A frigid and lifeless mountain-top? Or a teeming jungle? Through the ears of our hearts and minds, we can immediately transport ourselves anywhere we want to go through the power of our own imagination.

In the beginning . . .

But we're getting ahead of ourselves. Let's instead take a trip back in time, when home game machines and personal computers were young, and degenerate mutants pored over their misunderstood machines; a tragic metaphor that mirrored all too well their own social exile. They had a radical idea. They would make their own fun and have their precious machines play games.

As alluded to above, videogames are first and foremost about the visual component. The historical basis for this is that people interact with computers primarily through a visual interface. Beyond that, if an early computer had any sound capability at all, it was usually some form of "beep" or other notification tone. In the 1980s and 1990s, the first general-purpose PCs did not have the horsepower to do a lot of video processing, let alone play back any audio. They were not designed for multimedia presentations. Hence the need for a machine dedicated solely to playing video games – a game console. Because of this dependence on the hardware capabilities of the machine, there was a very close relationship between the advancement of games and the capabilities of the machines that ran them.

Early on, development tools and processes for game consoles were either non-existent, or very much tied to the particular machine itself. Artwork and graphics generated for one box could not be used on another system. The same went for sound. Basic art and music programs consisted of simple programs, assemblers and compilers, often written in some arcane and maniacally platform-specific language. Artists were constrained not only by the physical resources available to them, but also by their ability to write software. The code and the content on these machines were tightly linked.

Over time, content-creation tools became more mature. Two- and three-dimensional art packages began to emerge; digital audio editing and processing capabilities appeared on desktop computers; developers created more in-house tools, and viable in-house development libraries were written. As technology improved, it became easier to separate the art and music assets of a game from the software that displayed, or rendered, that content. As this happened we saw the rise of the game engine, coupled with homemade editors for content that went into that engine. Programmers created the engine, designers integrated the content elements, and together the two implemented the core logic and rules for the resulting title. Making a game wasn't only a programmer's job anymore.

In the mid-1990s, more and more cross-platform, integrated development environments and three-dimensional tools became available. Licensable engines gained ground, and in-house editors became polished programs, releasable to the public. In response to rapid content growth, middleware asset management software began to appear. The net result of these developments was that as games became more sophisticated, you could remain competitive by buying third-party solutions for those parts you didn't know how to write yourself.

The first computer games were nurtured in the hearts and minds of programmers with a good idea for a game story line, a lot of time and very little money. They started out modestly. The scene is very different today, where the development costs for premier game titles can easily run several million dollars. The combination of enhanced technology and the lure of larger profits is in a tug of war with rising production costs, increasing competition and shortening deadlines. This is forcing developers to launch products on all major platforms simultaneously in worldwide markets.

The Tao of now

The need to produce a cross-platform, simultaneous release requires precision coordinating of multiple assets, production talent and especially strong schedule milestone adherence. This is one of the core aspects driving the maturity of the industry value-chain. Better, more integrated content creation and integration tools, tighter management of development assets, and game engines with open-ended frameworks, re-usable for many different purposes – all of these combine to make it easier to port products across platforms while simultaneously localizing content for a coordinated, worldwide launch.

A developer's decision to release a title on multiple platforms to amortize development costs is a tricky proposition when it comes to audio. Because games history is what it is – fractured and limited by the emerging technology – the audio capabilities of each platform are inconsistent and different from one to another. Audio is being addressed in isolated, vendor-specific circumstances, which has led to proprietary and incompatible systems. There is no higher audio-centric consciousness at work to facilitate a consistent set of audio capabilities across the multitude of existing gaming platforms.

Game audio has always been, and will probably always be, in a constant state of change because the nature of technology is just that: ever-changing. The good news is that game players have matured and are demanding a higher quality experience that includes music and sound effects. The game development industry is paying more attention and devoting more monetary resources to audio each year. But many major problems still exist: audio still gets only 10% or less of the total RAM, storage and computer processing budget for the game. Most programmers working in audio do not have any formal training in music, sound design, audio recording or mixing and know little of the audio arts. Audio content creators know little about programming or how their audio content gets inserted into a game. And there are no high-level audio engines that allow a game's audio to be authored once and play back the same way on all platforms.

The audio systems found in contemporary game consoles and desktop computers aptly meet the basic technical challenges of delivering audio under computer control. These audio hardware and software devices can play back sound with good quality, and enjoy wide distribution due to the popularity of and market force that is computer games and entertainment software. Finding a solution to the problem of non-musical programmers and non-technical musicians requires education and recruitment. We need to find technically apt musicians and make them aware of the career opportunities in multimedia audio. The question of resources is a more difficult one to solve. Until it can be demonstrably shown that a game's financial success is directly related to the quality of its soundtrack, it will be hard to convince developers to give more resources over to audio. That's just the kind of ammunition we wish to present here.

What we're on about

In a game, the audio has to adapt and change logically to match the action on the screen. In all situations, the game player's decisions and actions determine what is seen and heard. This differs specifically from linear formats such as CDs or movies, where an audio artist knows exactly when a particular sound, piece of music or dialog is going to be played back and under what circumstances. What is needed here is an audio engine that can start and stop a sound, pause it, set its volume and pan, loop it, branch to another location in the same piece and transition from one piece of audio to the next. These are the fundamental building blocks necessary to create a clean and seamless interactive audio experience.

In addition, because games are shipped on multiple platforms we need to minimize platform-specific audio content creation and implementation by supporting a common and consistent set of higher-level audio behaviors across all platforms. This only makes sense from a business perspective. Both on the programming and content sides, development costs go down when the audio for a game can work across all platforms with minimal effort. Audio programmers spend less time writing platform specific code changes. Game composers and sound designers are more productive and effective when they are familiar with the audio integration process, and should be given the power to drive that integration without requiring any direct intervention from the programming team. All of this has a direct impact on how good the audio is in the final product.

Contemplating and planning a game's audio behavior and content takes a dedicated and concerted effort. Audio should be fully considered in the planning stages of a game, equal to the other art and video resources. This means that, from the start, the game's design and objectives must include audio as a component of programming and engineering. From the audio programmer's perspective, the sound engine must be efficient and reliable, and require minimal services and attention from the programming team and its leader. The sound artist should be able to easily and directly insert and tweak content, also with minimal programmer intervention. They bring their vision, or rather, audition to the game's design same as the video artist does. They should be treated with equal respect and given the resources and tools necessary to carry out that audition through into the final product. In this realizable dreamscape, new audio content does not require re-compilation. Cues that trigger an audio event are inserted into the game code once, where after the artist manages all changes to the audio via new content drops. In this way, audio artists not only deliver the raw content for a game, but are actively involved in its presentation and delivery throughout the game's development.

This book picks up the story of game audio in a period of significant transition. Present-day game machines have by and large freed us from the low-level concerns of getting sound out the back of the box. It is now our task to expose the creative potential these computer audio systems represent. The field of Computer Music grew from the desire to have technology serve musical purposes. In games, it is the other way around where all audio, in the form of music, dialog and sound effects, has become subservient to the technology. While a number of developers have put together in-house systems to address many of these problems, we hope to encourage a more public discussion of the Computer Music tradition in games via this book.

The Soundtrack Manager presented herein gives everyone the opportunity and means to creatively and musically fashion immersive, interactive, nonlinear auditory experiences in a game context. It provides the same functionality on whatever platform it runs, and puts the power of the computer audio systems of game platforms into the hands, ears, hearts and minds of the musicians and audio artists who use them.

2 Game audio APIs – Audio building blocks

As we saw in the previous chapter, there is a demanding legacy of programmatic efficiency and speed in the realm of games. Early computers and game systems had very constrained memory and storage resources. Coupled with limited display and media-handling capabilities, they presented many obstacles to overcome. The challenge of making fun and interesting games was as much about squeezing every last cycle of performance out of the machine as it was about the design of the game itself.

Don't call us, we'll call you

As few high-level development tools existed for early game systems, a programmer's only real choice was to get deep down into the system, and throw those bits around by hand. Thus is the genesis of the macho programmer image. Real men (admittedly, mostly young men and boys), making the hardware beg for mercy and trimming another 10 CPU cycles off their latest side-scrolling engine. Needless to say, the measure of a programmer's worth was in their ability to write in assembly language, living just above the silicon.

The early game programmer was the "Keeper of the Knowledge," and all game assets flowed through him. All waited anxiously for the programmer to descend from on-high (more like emerge from his dark and smelly cave) to give them an update on his progress, and to declare what assets he was going to add next. It can be argued that such centralized control was a good and necessary thing at the time. Computers were new, and things like integrated development environments, built-in file systems and memory resource managers did not exist. One had to know what the hardware did, and how to talk to it. A few, self-selecting group of resourceful and interested individuals dedicated themselves to understanding how the machine worked, and built all the necessary system components by hand to make a game come together. There was no other way.

Unfortunately, this concentration of power in the hands of the one, or maybe the few, has left a troubling legacy for current game development. Since each lead programmer feels that his is the only "right" way to do things, the majority of games are still re-written from the ground up every time. No self-respecting game programmer would ever use another's code. It is "crap" by definition, and must be completely re-done. This flawed development strategy hinders the development of new titles and slows progress on getting the game out on time.

To a game programmer, it's all about the code: how fast it runs, how efficiently it gets the job done, how cool the algorithm is. There is the fundamental belief that if you throw enough technology at a problem, it will go away and the end result will be great. I believe this faith in technology is misplaced. Games have evolved from the perspective that the most important thing is "how" you do a particular task, not "what" you want to do. Developers often get so caught up in the coolness of the code that they forget about what they were trying to achieve in the first place. Make no mistake – computer games use very sophisticated technologies that require skilled and knowledgeable programmers of all descriptions. However, the technology should serve the artistic needs of the game, not the other way around.

We're mad as heck

We, the audio professionals of the games industry, have been subservient to the whims of these Überprogrammers long enough. We are tired of having our content integrated into a game by people who have little or no musical skills and no idea of our artistic intentions. It is unacceptable to have to wait around for hours, possibly days, to hear the results of our latest submissions. We want to free ourselves from the tyranny of the game programmer bottleneck. We want to become the masters of our own destiny when it comes to integrating our audio art into a game. Furthermore, on whatever platform we find ourselves, we want to have a common set of actions and commands that will allow us to seamlessly and effectively create an interactive soundtrack in a game setting.

To understand where we want to go regarding interactive audio, we must first understand where we are. There are two distinct disciplines required to achieve great audio in a game. The first is composition – creating and assembling all the content that will be used in the game. The second is programming – designing and building a system to realize the audition of the composer in a game context. Once upon a time, in the pre-PC days, the same person often handled both of these tasks. But there are very few people who can perform both of these responsibilities well, and that's good! We can't all be experts in everything, and shouldn't try to be. We should let people do the jobs they're good at, and not set them, or our game, up for failure by asking them to do more than they know how to do. Thus, there should be (at least) two complementary audio professionals hired for each game: a composer and sound artist, and an audio programmer.

Ideally, teams of artists assemble all the audio content that goes into the game, from the background music tracks to the dialog and sound effects. All this material must be consistent with and support the overall artistic vision of the game. He or she should also have some knowledge of the technology that will into a game, but they should not be responsible for writing any code beyond some simple, and not-so-simple, scripts. (We'll talk more about this in Chapter 10.) The audio programmer, on the other hand, is responsible for writing the code that manages and plays back the audio content in an efficient and flexible manner. He or she must have an appreciation for what the composer is trying to create with their content, but should not be responsible for providing any of the audio content itself. It is their job to provide the necessary audio services for the game, understand how the audio software system fits into the larger structure of the game itself and to utilize all necessary platform systems and services.

Small game development houses typically do not have an in-house audio artist to make the content for their games, let alone a full-time staff audio programmer. I'm happy to say this is changing within the industry, especially in larger development shops. But in this all-too-common situation, audio is not well served. An audio artist is time and again not brought into the project until it's final phases, and has to create an entire body of work in a very short period of time. The programmer assigned to integrate that content is frequently pulled off some other part of the project to code the audio. He probably likes music and is interested in doing the work, but seldom has any previous experience or formal training in audio technology and signal processing, let alone even a basic understanding of music composition and performance. It is a credit to the legions of artists and programmers out there that we've gotten as far as we have. Nevertheless, we shouldn't have to settle for this nonoptimal situation anymore. With the tremendous growth of the games industry, game audio scoring and programming has come of age. It is now essential for a game to have great audio, and there are many dedicated professionals out there making it happen, with more in the pipeline. There's no excuse for bad audio in games anymore.

Third-party audio libraries

In this book, we concern ourselves primarily with audio programming for interactive games. It is our hope that by reading this book, people of all levels of responsibility and expertise in games will gain a greater appreciation and understanding of what it takes to make great game audio from the programming side. But if you're not an audio programmer, and don't happen to have any such creature in your immediate sphere of influence, what is to be done? Your game has to have audio. What options exist when you require audio programming support, but don't have the in-house expertise to make it happen?

Ten years ago, you couldn't buy a whitepaper on game audio. Today there are a growing number of technical game audio books out there, and a number of conferences and events with many dedicated game audio sessions. There are a few independent programming professionals targeting game audio out there in the marketplace, and the pool of trained audio programmers will almost certainly grow as the domain of games and multimedia entertainment grows.

That's all well and good, but you need a solution now, not months or years from now. Fortunately, there are a number of audio subsystems widely available today written explicitly for multimedia entertainment and gaming applications. These systems are incorporated into libraries that you can link into your game application either statically or dynamically at runtime. They make it very easy for a non-audio programmer to incorporate a fully functional audio system into a game without having to write one from the ground up. Ease of integration varies among these providers and from one platform to the next. There is always some amount of custom support or wrapper code to be written. However, the end result is that you can add audio support to your game almost immediately.

Game audio libraries come from many different sources: operating system manufacturers, three-dimensional audio companies, soundcard manufacturers, open-source audio communities and audio software companies. Their motivations and solutions, and the platforms they run on, vary depending on their business model, and that's OK. The most important factor you have to consider is, "On what platform(s) do I need audio support?" The answer to this question will define what options are available to you.

Beatnik Audio Engine

The Beatnik Audio Engine is a scalable software audio engine for PCs, personal digital assistants (PDAs), mobile phones and other digital devices. It provides a multi-timbral Musical Instrument Digital Interface (MIDI) wavetable synthesizer as well as a digital audio playback engine. It supports the simultaneous playback of multiple files types, and strives to produce uniform results across various devices. It has been proven in a number of PC and console games, and is the standard audio solution for Sun Microsystem's J2SE platform.

FMOD

One of the more recent additions to the third-party audio fray is the FMOD music and sound effects system from Firelight Technologies. It is a cross-platform audio engine that runs on a wide variety of platforms including Windows, Windows CE, Linux, the Macintosh, and now on the GameCube, PlayStation2 and the XBox game consoles. It supports the DirectX Audio API on the PC, but does not require it. It will work on any machine with a 16-bit sound card. It uses the built-in Sound Manager on the Macintosh, and supports both Open Sound System and Advanced Linux Sound Architecture under Linux (see below).

DirectX Audio

One of the most widely distributed audio libraries is the DirectX Audio Application Programming Interface (API) from Microsoft. It is part of their larger DirectX family of technologies, and is available exclusively on both the various flavors of the Windows OS and the XBox game console. Each of the DirectX APIs functions as a kind of bridge for the hardware and the software to "talk" to one another. DirectX Audio gives applications access to the advanced features of high-performance sound cards on a given machine, and controls the low-level functions of sound mixing and output. Initially written for the PC, DirectX has been completely re-written for the Xbox to take full advantage of a slew of dedicated audio hardware available on that platform.

GameCODA, Miles Sound System and MusyX

Other commercially available audio libraries are the GameCODA system from Sensaura, the Miles Sound System from RAD Game Tools, and the MusyX sound effects and music system from Factor 5. The GameCODA system runs on the PC, PlayStation2, XBox and GameCube, while the Miles Sound System runs on the PC and Macintosh platforms. MusyX is the official sound and music system for the Nintendo family of game platforms. It features an integrated editing tool that emulates the target platform using a couple of PCs, and contains a proprietary programming language, called SMaL (for Sound Macro Language) to develop, design and control sound effects and music instruments. Unfortunately, none of these commercial solutions are freely downloadable, and are only available to licensed developers.

Mac OS X Core Audio

The Carbon Sound Manager in OS X is an efficient and flexible audio solution for doing all manner of digital audio handling and signal processing through what they call AudioUnits. Core Audio handles all data as 32-bit floating-point numbers up to 24bit/96 kHz, and provides native multi-channel audio scalable to n-channels. It also provides low-latency MIDI I/O services through its CoreMIDI framework API that allows you to route MIDI data between in, out and across your applications. The API also provides extensive support for MIDI stream and configuration management in a multi-port setting.

UNIX solutions

On UNIX platforms, vendors traditionally provide their own APIs for processing digital audio. This means that applications written to a particular UNIX audio API often have to be re-written or ported, with possible loss of functionality, to other versions of UNIX. These proprietary APIs didn't always provide the same broad range of functionality available on the PC or Macintosh platforms, either. With the advent of streaming audio, speech recognition and generation, computer telephony, Java and other multimedia technologies, applications on UNIX can now provide the same audio capabilities as those found on the Windows and Macintosh operating systems through the Open Sound System (OSS). OSS is a set of device drivers that provide a uniform API across all the major UNIX architectures, and provides synchronized audio capability for desktop video and animation playback. Applications written to the OSS API need to be designed once and then simply re-compiled on any supported UNIX architecture.

The Advanced Linux Sound Architecture (ALSA) provides audio and MIDI functionality for the Linux operating system. ALSA supports all types of audio interfaces, from consumer soundcards to professional multichannel audio interfaces. It has some distinct advantages over OSS: it is thread safe, provides user-space software mixing, merges multiple sound cards into a single virtual device, and provides a consistent and flexible mixer architecture.

OpenAL

The Open Audio Library (OpenAL) is a cross-platform three-dimensional audio API available for the PC and Macintosh platforms. Its primary purpose is to allow a programmer to position audio sources in three-dimensional space around a listener. The OpenAL software development kit (SDK) is available from the Creative Labs developer site, and was co-developed by Creative Labs and Loki Entertainment Software.

All of the above-mentioned audio libraries provide low-level control of and access to an application's audio resources. They supply a common set of audio functionality and facility across a number of different platforms. To varying degrees, they define and describe a vocabulary of what I like to call low-level audio "operators." Just as the arithmetic operators $+$, $-$, $*$ and $/$ are used to solve larger mathematical problems, the functions in these APIs can be strung together in very specific ways to perform larger audio tasks. The bottom line is that these libraries can provide the audio control and handling necessary to build a functional game audio system.

Your mission, should you choose to accept it

As in many programming endeavors, the devil is in the details. As the audio programmer, you have a lot of details to keep straight. For example, you have to initialize the platform in question to handle the audio in your game. If the library runs on many different platforms, you have to explicitly select the desired platform, as well as the specific output device, soundcard or driver attached to that system. All of the audio will be mixed together before being played out, so you have to decide what type of mixer best suits your purposes: interpolating for high quality, or truncating for lower quality but greater speed. How big should the mixing buffer be? A small buffer provides low latency between the mixing of the sounds and when they appear at the output, but a larger buffer doesn't have to be tended to as often. Some game platforms provide hardware-accelerated audio channels for faster mixes or enhanced three-dimensional sound capabilities. How many hardware channels will you need for your game? Can you afford the processor- and quality-hit when you fall back to software rendering when you run out of hardware channels? Sounds have to be loaded in and out of memory throughout a game. Will you supply your own memory management system or utilize the internal calls of the platform? All of these decisions have to be made just to initialize your library of choice.

Before you play back a single sample, you also have to set the audio output configuration. Should the game use stereo speakers, headphones or a multichannel setup? There is no way to know this a priori, so you have to accommodate everything. Will you be using 8- or 16-bit data? Are your sounds mono or stereo, or some of each? Is the audio stored signed or unsigned? What is the endian-ness of your data? Does it match that of the platform? What audio formats does your engine support (WAV, AIF, SD2, MP3, ADPCM, etc.)? What audio formats does the platform support? What audio codecs are available, and will you be using them? What is your output sampling rate: 8 kHz, 11 k, 22 k, 44.1 k? Will you have access to or be using a built-in CD player?

Only once all those decisions have been made can audio data be loaded and played. If your sounds are short enough to fit into memory all at once, that makes things easy. Larger files, such as lengthy dialog background tracks, will have to be streamed off the hard disk a little at a time. Do any of the sounds have to loop? How many times? Can you set up any notifications to tell you when a sound has looped, reached a certain point or stopped playing? You will need to set the volume and pan of all your sounds, and maybe change their pitch or spatial location. Will you be playing back MIDI or other song files? How many notes can you play simultaneously? Can you use custom digital audio sound banks with MIDI, and if so, are there any special or proprietary tools you need to make them? What happens when you run out of available voices? Can you prioritize your sounds? Does the particular API you're using address these issues?

The list goes on and on. Fortunately, each of the above items (and many more!) are likely addressed by some function in your third-party audio API of choice. They do a good job of handling the nuts and bolts of playing back audio under computer control. However, the biggest drawback about these APIs is that they do not define what we want an audio system to do from a creative, musical standpoint. To do that, and to put the reins of control into the audio artist's hands, we have to abstract the behavior of the audio system from the underlying system operations.

Creative thoughts

When you listen to a piece of music, either live or from a recording, you don't think of the mechanics involved in playing the various instruments you're hearing. A composer, on the other hand, needs to be aware of the physical ranges and capabilities of the instrument before she writes music for it. Otherwise she runs the risk of writing a part the intended instrument cannot play. She doesn't have to concern herself with the distribution of the parts for the rehearsal or the resilience of the clarinetist's reed, or the amount of rosin on the bows of the first violins. All those details, while absolutely necessary to perform the music, are not the language of musical, artistic expression.

Similarly, the goal of audio in a game is to envelop the listener and reinforce the action via a coordinated auditory sensory channel. If done well, this further draws the player into the game. But the same person who writes the audio content shouldn't have to become a proficient programmer to realize their audition. It's a matter of letting people do the jobs they know how to do, and leaving the rest of the task to others. You generally don't want programmers writing your music. Why make composers write software? That ability is not a requirement to create good game audio content.

The audio programmer is not so concerned with the audio content as he is with playing everything at the right time and place. From the audio content provider's perspective, programming has a direct effect on how her content sounds, but she has no direct control over it. The game programmer doesn't want to be bothered about the audio, until it starts to impact the game's visual frame rate or introduce bugs. And finally, so long as there's a great soundtrack, the game player couldn't care less about any of this stuff.

So where's the happy medium? What can we do to accommodate these seemingly incompatible and competitive concerns regarding game audio?

The task at hand

In the chapters ahead, we develop an interactive game audio engine that is independent of the platform on which it runs. It abstracts the use of audio in games both from a programming and artistic perspective. The high-level musical expression of the composer guides the underlying, algorithmic implementation, and approaches the problem more from a "what" rather than a "how" point of view. At this level of abstraction, any game development company can take these concepts and apply them to their own development environment, or a specific vendor's game system. As a matter of fact, the Soundtrack Manager documented here can easily be placed on top of any of the low-level audio APIs mentioned in this chapter.

Hail, fair game audio people! From this moment on, we claim and shape our own destiny.

3 Introduction to interactive game audio

What now?

A game composer should have the ability to construct an appropriate soundtrack for the game. In the case of movies, this is a relatively straightforward task, as the composer knows exactly when and where certain events are going to happen. This is because film is a linear medium, experienced from start to finish upon each viewing and never changing. Games are a totally different paradigm. Here the composer has limited knowledge of when or where things will happen, as that's all under the control of the player. It's a different way of thinking about music and flow, and must be accommodated. Otherwise, you can wind up in an endless scene with a looping background track that quickly becomes monotonous and boring.

Thing 1 and thing 2

To avoid this, our first goal is to design and build an interactive audio system that allows the audio artist to construct a compelling soundtrack that will adapt and change in response to the player's actions in real-time and in an appropriate manner for the current game situation. They should also have to author that content only once for all the platforms on which the game will run. Therefore, our second goal is to separate the desired operation and behaviors of an interactive audio system from its platform-specific implementation.

This is a pretty tall order. From a programming perspective, how do we accomplish this?

The simple, yet powerful, answer is to first approach the problem from an artistic, not a technical, perspective. We turn the conventional thinking about a game audio system on its head by considering "what" we want to do before we delve into the "how" of making it happen. We must define and then support the fundamental actions necessary to afford a composer or sound designer the flexibility they desire when constructing an interactive soundtrack. On the most basic level, our audio software system has to play, stop, pause and resume whatever content we throw at it. It also has to loop, unloop, mute and unmute sounds, set their volume and pan positions, and fade sounds in or out. Most, if not all, of these operations are supported to some degree in the APIs we examined in the previous chapter. But we want to do it all, and without having to use a different syntax on every platform.

Beyond these basic operations, however, the composer needs the ability to assemble play lists of different audio material. Using these constructs, composers can specify and direct the flow of the audio from one piece to the next. They must have the facility to create, add, remove or clear these lists and have the means to jump or segue to any piece of audio content within them. They should be able to directly address and manipulate any subcomponents or tracks of the audio content, and have some signaling or callback mechanism to keep track of the sounds while they're playing and when they finish. There is precious little in the audio APIs we looked at in the previous chapter that offers this kind of functionality "out of the box." One has to build these behaviors on top of what they provide. One could argue that many game developers have done this before, so this is not new. I would agree with this assertion, but only to a point. The higher-level musical systems that use these underlying APIs have not been sufficiently disseminated to the game community at large, but rather reside in the proprietary hands and libraries of many different individual companies. The Soundtrack Manager is an attempt to bring the ideas and details of one interactive audio system into the light of day for everyone's use and discussion.

Audio supports visual media in a primal and visceral way. In a game, the coordinated stimulation of the aural and visual senses creates a synergy that makes a game more engrossing, enveloping and interesting. But to do this well, the audio in a game must respond logically and gracefully to current game conditions. It should draw you into the game and suspend your disbelief in what you're seeing and hearing. It should "sound natural," and not violate your aural expectations of the scene or surrounding environment. We should be able to associate sounds with specific characters or objects in a game, and those sounds should be synchronized with the visual rendering of that object. All of this requires well-defined lines of communication between the sound engine and the game application itself. And no matter what the underlying audio resource, these behaviors must be addressed and work the same across all platforms.

Finally, a record of what audio is currently playing in the game should be preserved so it can be restarted across different sessions. Therefore, we must also provide commands to pause, resume, reset, save and reload a game's audio configuration.

No party affiliation

As mentioned above, a principal objective of our audio system is that it be platform-independent. Everything we do should be described and coded at a level above the platform itself. This is analogous to writing an orchestral score. In this example, a composer writes down the musical instructions for all the instruments they desire in a sufficiently accurate way so that the parts can be played back by some group of musicians at a later point in time. The composer generally does not have any a priori knowledge of the specific instrumentalists or conductors who will be performing his or her opus, beyond it's premiere at least. The music is written to the idea of an orchestra, with an understanding of what each instrument does and how it sounds.

Only when the score is played will the notes and other written gestures be "rendered" into the acoustic wave patterns we perceive as music. It is also hoped that different performances of the piece using different performers will sound similar to one other and to what the composer intended. If it didn't, our perception of music would be radically different and much more chaotic than it is. Imagine going to hear a rock concert only to have a folk revival break out. But this is exactly the kind of thing that tends to happen when a different person on different platforms integrates a sound artist's content into a game. The basic understanding of how the content goes together is lost, and all manner of problems ensue.

Continuing on with the orchestral score analogy, once the score is finished the orchestra contracted to play it should not be allowed to change any of the parts. Imagine the cacophony and anarchy of a "do-it-yourself" Beethoven's Ninth Symphony. An amusing thought, but it doesn't reflect how Ludwig heard it (albeit only in his head). The operative point here is that the performers of a score, or piece of software, shouldn't be allowed to make changes to one's opus, at least not without your permission. We could rely on everyone's good nature and pledge not to muck around with your stuff. The reality of the situation, especially in software, is that you have to explicitly protect yourself against this kind of misuse. Therefore, as you'll see below, we install gates and put up roadblocks in our software to deter any over-enthusiastic developers from making any unauthorized changes or taking extraordinary liberties with our code. Wherever possible, we wrap up our code to the extent that an external user won't ever know what's there. Sure there're ways around it, but the boundaries have to be consciously breached to do so.

Each orchestral score requires a specific instrumentation. Hiring musicians costs a lot of money. If the score only requires 25 players, there's no need to hire a 100-piece orchestra. In game audio, this is analogous to requesting only those services we require for our specific purposes. As far as the software is concerned, we want our sound library to be only as simple as it needs to be, and no simpler. There's no need to include support for something if it's not going to be used. This requires the ability to reconfigure the resources necessary for any one particular product.

Into the code

The code for all the libraries and applications presented in this book is included on the accompanying CD. The root Code folder contains three separate subfolders: Apps, Libs and Sound. The Apps folder contains the code for two very important applications known as MIDI2NMD and SNDEVENT. MIDI2NMD is covered in great detail in Chapter 6, and is a utility to transform a Standard MIDI file into a new form that allows for efficient mid-file startup, seeking, looping and branching capabilities. SNDEVENT, described in Chapter 10, is an application used by audio artists to compile English-language text scripts that give them full control over Soundtrack Manager operations without having to be a programmer.

The Sound folder is where we keep all the code for the three different versions of the Soundtrack Manager library we supply on the CD. All of these versions run under the Windows operating system. The core, platform-agnostic portion of the sound engine code resides in this top `Sound` folder. Under this folder is the `WIN9X` folder. The files in this directory are common to all Windows versions of the Soundtrack Manager. Under here we have the `DSMIX` folder. This is where we keep those versions of our files necessary to directly use the audio mixing services of DirectX. Also under the `WIN9X` folder is the `WIN9XMIX` folder. This folder contains those versions of our files necessary to do the audio mixing on our own outside of DirectX. There are two versions of the non-DirectX Soundtrack Manager provided on the CD, reflected in the subfolders of `FP` and `MMX`. These folders contain those versions of our files for doing the mixing using either floating-point assembly instructions, or the MMX instruction set. You will notice that the different directories contain files of the same names. This is so we can maintain similar-looking projects for each of the different versions of our library. However, the published API in each of the header files in the top Sound directory does not change across versions. Only the internal contents of the functions themselves change among the three versions.

Three Visual C++ projects are included on the CD, corresponding to the three versions of the Soundtrack Manager. These can be found in the `DSMIX`, `FPMIX` and `MMXMIX` folders inside the `Sound` folder.

The `Libs` folder is where we place the various `Debug` and `Release` versions of the Soundtrack Manager library: `SLIBDS`, `SLIBFP` and `SLIBMMX`. This is also where we keep the source files necessary for the utility libraries included in the `SNDEVENT` application: `Xbase`, `Xparse` and `Xlnparse`.

Modules

The `modules.h` header file (in the `\Code\Sound` folder) is one of the most important structural elements in our library. The Soundtrack Manager is comprised of multiple functional components, which we call modules. In this header file, we customize the resources and functionality included in a specific version of the library by using several preprocessor directives. For example, to include the ability to play MIDI files, we define the Soundtrack Manager symbol `__SLIB_MODULE_MIDIFILE__` via the preprocessor directive: `#define __SLIB_MODULE_MIDIFILE__ 1`. If we want to include support for digital audio playback, we issue the `#define __SLIB_MODULE_ SAMPLE__ 1` directive. Just about every source file in our library includes this header file. In other parts of our code, when we come to a section that deals specifically with MIDI files or digital audio playback, we can wrap that code with a test to see if we need to include those statements. For example, to include a section of code dealing with MIDI file playback, we bracket the code inside a conditional preprocessor `if` block:

```
#if __SLIB_MODULE_MIDIFILE__
  //lines of code dealing with MIDI file playback here
#endif
```

If `__SLIB_MODULE_MIDIFILE__` is not defined, that code is ignored. In this way, we can conditionally compile library support for those modules and services we choose. This also serves to limit the size of our library if we don't require all the modules.

Several of the modules also contain sub-module definitions. These definitions let us refine the capabilities we build into a particular version of the sound engine. We can therefore build different versions of the sound library for use with different applications as is necessary and prudent. For instance, under the `__SLIB_MODULE_SAMPLE__` directive in `SM.h`, we have the ability to support numerous digital audio data formats. Defining `__SLIB_SUBMODULE_AIFF__` will allow us to read AIF files, while `__SLIB_SUBMODULE_ADPCM__` will add Adaptive Delta Pulse Code Modulation (ADPCM) decompression support to the library. We can also add new module definitions as we choose in the future, simply by adding more such directives. The list of possible modules and submodules is shown below:

```
__SLIB_MODULE_MIDIFILE__
    __SLIB_SUBMODULE_PLAYLIST__
__SLIB_MODULE_PROXIMITY_SOUND__
__SLIB_MODULE_SCRIPT__
__SLIB_MODULE_SAMPLE__
    __SLIB_SUBMODULE_AIFF__
    __SLIB_SUBMODULE_WAV__
    __SLIB_SUBMODULE_ADPCM__
    __SLIB_SUBMODULE_RAW__
    __SLIB_SUBMODULE_RAM__
__SLIB_MODULE_CDAUDIO__
```

We will see as we go along what each of these modules do, and how they are accommodated in the code.

Devices

In the Soundtrack Manager, there is also the concept of a `device`. A `device` is that part of the platform providing some input or output service. In a similar way as we did with modules above, we can customize the devices we support through a different set of preprocessor directives. Also inside the `modules.h` file, the statement `#define __SLIB_DEVICE_MIDIIN__`, for example, will include the code for reading and responding to MIDI messages coming in from the outside world. This is very useful when testing our MIDI synthesizer engine (which we'll cover later in Chapter 7). We can send known MIDI messages to our engine from an external keyboard controller or sequencer application and observe its behavior. When it comes time to ship the game, we remove this MIDI input processing capability from the final product simply by commenting out the `__SLIB_DEVICE_MIDIIN__` definition and rebuild. All the supporting code for that feature disappears with a few simple keystrokes in one single file. To extend our capabilities in the future, new device definitions can be added. The support code for that new device would then be wrapped in new conditional if blocks. The list of possible Soundtrack Manager `devices` is shown below:

```
__SLIB_DEVICE_PCM__
__SLIB_DEVICE_MIDIOUT__
__SLIB_DEVICE_MIDIIN__
```

Our first source file

Let's take a look at one source file and see specifically how we accomplish each of these boundaries and layers. We begin with the DSMIX version of the sound library. You'll find the code listing for SM.c in the \Code\Sound folder on the CD.

The first thing we see is the preprocessor directive #define __SLIB__. This is our access control statement. For those of you who may be familiar with the C++ programming language, this works in a similar way as the public keyword inside a C++ class definition. The name __SLIB__ is the identifier we use to expose our sound libraries core functionality and interfaces. When it is defined, we're as naked as the Emperor in his new clothes. Every single variable, function call or data structure throughout our library is visible. However, when it is not defined, it selectively hides those components we don't want the outside world to see. This is done so other applications can include sound library header files but only see the interface and variables we wish them to see. Of course, an external module could include this directive, but this would be a severe violation of the code-writing rules of etiquette, if not our general trust. The presence of this directive means that the following is for internal use only.

The next thing we do is to include the header file, modules.h (in the \Code \Sound folder). As we mentioned above, the Soundtrack Manager is comprised of multiple functional components, or modules. By including this file, we include only that code necessary to support what we're after.

Continuing along in SM.c, you'll notice a number of other header files are included that specify the additional services and definitions we wish to incorporate. However, a single header file may be included only once, or else we end up with multiple definitions of things. It is a tricky proposition to accommodate this restriction, as it is common for several source files to require the definitions of a single header file. Looking at SM.h (also in the \Code\Sound folder), we define a unique symbol for each header file and use the #ifndef directive to check if the symbol for that header file has already been defined. If it hasn't, we define the symbol in the very next statement, and continue processing the header statements until we encounter the next #endif statement (conveniently placed at the end of this header file):

```
#ifndef __SM_H__
#define __SM_H__
    . . .
#endif
```

For any subsequent files that include SM.h, the compiler will see that the symbol __SM_H__ has already been defined, and not include that header information again.

This mechanism assures us that the definitions and services our current source file requires will be included, and we won't get any nasty compiler errors for multiply defined symbols or illegal redefinitions.

Types

Just about every header file in our library includes `stypes.h` (in the `\Code\Sound` directory). It is here that we specify the data types and names we're going to use throughout our library, along with a number of useful macros. Programmers all have a favorite compiler they use (or are required to use) to build their games. One compiler may define an unsigned, 16-bit integer as a `uint16`, while another compiler might simply call it an `int`. Because we can never tell what compiler will be used to build our sound library, we define and use our own terms for all standard data types in `stypes.h`. This prevents data type errors and conflicts.

The majority of the data types shown in the `stypes.h` header file are self-explanatory, but there are a few data types that require some additional explanation. In the Soundtrack Manager, we make liberal use of fixed-point integers. There are no floating-point operations anywhere in the code. This is done primarily for speed, as floating-point instructions are more expensive, that is, take more cycles, than their integer counterparts. Secondarily, some platforms do not support floating-point operations. Therefore, to run on all platforms, we use integers exclusively.

Fixed-point integers allow us to represent numbers less than one. The value of any bit is normally the number 2 raised to the power of its position. For example, the first bit, or bit 0, has a value of (2^0) = 1. The second bit's value is calculated to be (2^1) = 2, and so on up to however many bits there are in the digital word. Numbers less than one are represented by "fractional bits." The values of these bits are found by raising 2 to a negative power. In the `FIXED4_12` Soundtrack Manager data type, the high 4-bits of this 16-bit word are treated as a traditional integer quantity. We shift the point at which we start our positive exponent count up to Bit 12 = (2^0) = 1. Bit 13 = (2^1) = 2, and so on up to bit 15 = (2^3) = 8. The low 12-bits are treated as a fraction, starting with bit 11 that has a value of (2^−1) = 0.5. Bit 10 = (2^−2) = 0.25, and so on down to bit 0 which = (2^−12) = 0.000244140625. Setting all the bits in a `FIXED4_12` number, or `0xFFFF`, is the number 15.999755859375. The `FIXED20_12` data type is a 32-bit fixed-point integer: a 20-bit number with 12-fractional bits. The maximum value of this number is `0xFFFFFFFF`, or 1048575.999755859375.

Fixed-point numbers give us a way of expressing fractional numbers with high precision while still using integers. Then we just lop off their fractional parts when we have to use the result. Even this lopping is easy. There's no division necessary; just a simple right-shift by 12 gives us the number we want.

Platforms and options

The next file we include in `SM.h` is the all-important `SMPlatfm.h` (in the `\Code\Sound\Win9x` directory). Together with the `SMPlatfm.c` source file, this is where we take care of all the platform-specific information, data structures, definitions and function prototypes we'll need to run the Soundtrack Manager on our platform of choice. Anything that needs to be done on a platform-specific level is concentrated here: things like memory management, timer sources and interrupts, specific variable information, activation and de-activation, and critical sections. It's all localized here. To tailor the library for a different platform, we write new versions of the SMPlatfm.c and SMPlatfm.h files, and re-compile.

SM.h continues with the type definition of a public SMOPTIONS structure that stores some basic information about the configuration of our sound library. It's public because it is not enclosed in any __SLIB__ conditional statements:

```
typedef struct _smoptions {
  void (*pfnErrorPrint)();
  struct _platforminfo platformInfo;
} SMOPTIONS;
```

pfnErrorPrint is a function pointer that allows us to specify an application error reporting function. platformInfo is a platform-specific information structure that encapsulates each platform's necessary information. In the DSMIX version we're looking at, this platformInfo structure is defined in SMPlatfm.h (again in the \Code\Sound\Win9x directory). Following SMOPTIONS is a private, __SLIB__-enclosed SMSTATUS structure:

```
typedef struct _smstatus {
  volatile   UINT32        frame;
  volatile   FIXED20_12    frameRate;
  INT16      PCMDevice;
  INT16      MIDIDevice;
  INT16      MIDIInDevice;
  UINT16     numDevices;
} SMSTATUS;
```

Again, the name __SLIB__ is the identifier we use to expose our sound libraries core functionality and interfaces. The SMSTATUS structure informs the internal Soundtrack Manager routines about the individual devices we are using and the current frame count of the engine (see below).

Frame-up

Each timer interrupt in the Soundtrack Manager is called a "frame," and frame keeps track of the current number of timer interrupts. frameRate is the inverse of the timer period, and is used in the tempo calculations for MIDI file playback (see Chapter 5). We will see in the next chapter that both digital audio and MIDI require some system-level services to do their thing. We keep track of these platform-specific resources in an array called Devices, and index that array using the PCMDevice,

MIDIDevice and MIDIInDevice members of this structure. numDevices is simply the total number of devices we are handling on this platform.

The public interface for the Soundtrack Manager's core module is listed next. These functions are called by an external application to start up, shut down, suspend or resume our library's execution. One of the most important functions is the SMSetSYSStringHandler method, which tells the Soundtrack Manager the name of the routine to call to print internal Soundtrack Manager error strings. This link is essential when developing and debugging our code, and can be disabled just before release by defining the SM_NO_ERROR_MESSAGES symbol before final compilation. The final lines of SM.h list a number of private, internal functions and global variables.

Updating modules

There are many operations we have to perform over the course of our sound engine's life. On start up we have to set up a number of internal structures, variables and/or arrays for each of the services we've chosen to include. When the application exits, we have to cleanly shut everything down and free all the memory we have used. During gameplay, we have to manage the playback of all the different sounds, adjust their volume or pan position, start or stop sounds, load and unload resources, and maintain our queues or play lists, all in response to what is happening in the game. When the game is paused and resumed, we have to perform different management tasks so that the audio stops and restarts as appropriate for each audible component of our soundtrack. These various tasks are organized into a set of logical operations or procedures for each of the modules, or functional components, we defined earlier in this chapter. All module-specific operations are grouped into an array of function pointers called ModuleProcs in SM.c. For each of the defined modules, we construct a list of the addresses of the functions we'll call to initialize, activate, deactivate, update and uninitialize that resource. We are assured that only the addresses of those functions pertaining to our included modules are placed in the ModuleProcs array via the specifically defined modules discussed above.

Start me up

At last we come to some real executable code. The SMInit function in SM.c is the first routine a game must call to start up the sound engine. In the first few lines of this function, we initialize our memory management system, all of our volume-handling code and all of our desired devices:

```
if(InitMemMgr()) {
  if(InitVolumes()) {
    if(InitDevices()) {
```

Next, we initialize each of the modules we've included in our library from the first entry in our `ModuleProcs` array. If this is successful, we proceed to add that modules' procedures to a set of global PROCLISTs we will use to administer to our selected modules at various points throughout the game. A PROCLIST is nothing more than a dynamically allocated list of function pointers:

```
typedef struct _proclist {
  void* proc;
  struct _proclist *next;
} PROCLIST;
```

In the code below, we build the global `ExitProcList`, `ActivateProcList`, `DeactivateProcList` and `UpdateProcList`:

```
i = 0;
while(moduleInitProc = ModuleProcs[i][0]) {
    //Execute initialization proc
    if(status && moduleInitProc()) {
      if(moduleUninitProc = ModuleProcs[i][1]) {
        ExitProcList = ProcListAddHead(ExitProcList,
                       moduleUninitProc);
      }
      if(moduleActivateProc = ModuleProcs[i][2]) {
        ActivateProcList =
          ProcListAddTail(ActivateProcList,
            (void*)moduleActivateProc);
      }
      if(moduleDeactivateProc = ModuleProcs[i][3]) {
        DeactivateProcList =
          ProcListAddHead(DeactivateProcList,
            (void*)moduleDeactivateProc);
      }
      if(moduleUpdateProc = ModuleProcs[i][4]) {
        UpdateProcList =
          ProcListAddHead(UpdateProcList,
            moduleUpdateProc);
      }
    } else {
      status = FALSE;
    }
    i++;
  }
```

If all goes as planned, we perform any necessary platform-specific initialization, and we're off:

```
if(status && InitPlatform()) {
  SMInfo.frame = 0L;
  SMInfo.frameRate = PlatformGetFrameRate();
  return(TRUE);
}
```

But what if things don't go as planned? Easy! In that case, we skip initializing the platform, and shut down our library. For any successfully initialized module, we have a list of what to do to uninitialize that module in our global PROCLISTs. We call each of them in turn, and we're outta there:

```
        ProcListExecute(ExitProcList);
        ProcListFree(ExitProcList);
        ProcListFree(UpdateProcList);
        ProcListFree(ActivateProcList);
        ProcListFree(DeactivateProcList);
        UninitDevices();
      }
    UninitVolumes();
  }
  UninitMemMgr();
}
return(FALSE);
```

Initializing our library, just to uninitialize it and shut it down, doesn't seem very sporting, or useful. We've made sure we don't leave any resources hanging and there aren't any memory leaks, which is important. But the more interesting behavior is what happens after the successful initialization of the library.

It's about time

The perception of music is very sensitive to temporal order. Our tolerance for miscued or late events is much greater for visual stimuli than it is for auditory events. We can drop frames of a movie or animation and never be the wiser. But drop a few notes, and it's glaringly obvious. Therefore, it is imperative that we keep our audio system moving along, presenting all the notes and sound effects and musical passages in a timely fashion. To accommodate this human auditory acuity, we must use a high-resolution timer to govern the temporal presentation of all our sounds. As you might expect, the mechanisms for creating or getting access to such a timer are different from platform to platform. Never fear. In SM.c, after we've successfully initialized all of our desired resources, we call InitPlatform (in `\Code\Sound\Win9x\SMPlatfm.c`, of course), to set up this timer. This function performs all the necessary, platform-specific operations to set up a steady, repetitive timer interrupt. Each time our high-resolution timer period expires, a private, internal function, `PlatformTimeProc` (again in `SMPlatfm.c`), will be called. This in turn calls our platform-agnostic update routine, `SoundManager` (in `\Code\Sound\SM.c`) that is the core of our engine.

I'd like to speak to the manager

While we're here, let's take a closer look at what our periodic `SoundManager` routine actually does.

First, we increment the `SMSTATUS` frame counter. This is simply a count of how many times our interrupt has fired since the library was initialized. We next update all of our defined `devices`, update the volumes for all playing resources, and call each of our update routines through the procedure list we set up during initialization. Finally, we perform any platform-specific operations that may be necessary, and exit:

```
do {
  //Continue updating based on mixer progress
  ok = UpdateDevices();

  //Update all modules
  UpdateVolumes();
  ProcListExecute(UpdateProcList);
  UpdatePlatform();
} while (ok);
```

The amount of work to do in each of these procedure lists or update routines depends entirely on what audio is playing at any given time. But this `SoundManager` routine encapsulates all of those operations into a few calls.

Leaving the scene

The function `SMUninit` (in `\Code\Sound\SM.c`) completes our survey of the Soundtrack Manager's functional organization. This routine deactivates the library by calling `SMDeactivate`. All currently sounding resources, be they MIDI notes or Pulse Code Modulation (PCM) samples or the CD player, are turned off and all the audio devices and resources we claimed when we first activated our library are released. Next, we uninitialize all our modules by executing the list of exit procedures. We don't have to worry about or check what modules we've been using, as these were selected via the various preprocessor definitions in `modules.h`. We complete the shutdown of our library by uninitializing our platform, devices, volume controls and memory manager. These Soundtrack Manager components are uninitialized in the reverse order to their initialization. This logical shut down and release of our library and its resources ensures we're not left with any memory leaks or orphaned resources. We've left the campsite as clean as we first found it, and packed out our trash.

Summary

In this chapter, we've presented the two major goals of this audio library: to design and build an interactive audio system that allows the audio artist to construct a compelling soundtrack, and to separate the desired operation and behaviors of an interactive audio system from its platform-specific implementation. We approach the design of this game audio system by first considering "what" we want to do before we delve into the "how" of making it happen. We presented an introduction to the structure and operation of the code, and some of the language mechanisms we use to support that organization.

You have also seen in this chapter how the code within a project is set up to accommodate different services and resources. The same kind of thing is done for the source files of the various projects: we segregate the code into discrete folders for specific platforms or capabilities. The organization and modular design of our library, both within the source code itself and among the files and folders themselves, makes it easy to extend and change the library's functionality and to target specific platforms.

CD-ness

A few final words on the organization of the source code found on the accompanying CD. In this book, we initially present code to use the DirectSound interface on the PC to mix our sounds together using the DSMIX Visual C++ project. We have also provided two other Visual C++ projects in the FPMIX and MMXMIX folders. As you will see, these projects all contain the same collection of files (with only a few exceptions). In these last two projects, we will show how to bypass DirectSound and do all our mixing using a custom software wavetable synthesizer. FPMIX will use a mixer written in floating-point assembly code. MMXMIX shows how to take advantage of a particular platform's capabilities by using the MMX instruction set to speed up our audio mixing.

4 Digital audio services – open up, it's the DA!

This chapter focuses on one game audio media type – digital audio (DA). A specific group of high-level audio operations pertaining to DA are developed including access and control, timers, update rates and system services. Platform-specific operations are also compartmentalized for easy replacing. All of this is in line with the goals of the Soundtrack Manager which is to support digital audio playback in an interactive and musical fashion. This fundamental perspective drives everything we do and provide for in our DA data structures and API.

Now you see it, now you don't

Digital audio results when a continuously variable analog signal is electronically transformed into a discrete, multi-level digital signal. A sample-and-hold circuit measures, or samples, the instantaneous voltage, or amplitude, of an analog audio signal and holds that value until an analog-to-digital converter (ADC) converts it into a binary number. The sample-and-hold circuit then reads the next instantaneous amplitude and holds it for the ADC. This occurs many times per second as the signal's alternating voltage rises and falls. As a result, the smoothly varying analog waveform is converted into a series of "stair steps." "Snapshots" of the input signal's voltage are taken thousands of times per second via an ADC and saved as discrete samples on some storage medium. The rate at which the instantaneous-amplitude measurements are taken is called the sampling rate, and the time between measurements is called the sampling period. The more often measurements are taken, the higher the frequency that can be accurately represented. In visual terms, this is akin to a strobe light on a dance floor. We see only the instantaneous positions of the crowd during the flash, and see nothing between flashes. This is called "quantization." If we speed up the flash rate of the strobe, we get a pretty good sense of the continuous nature of the dancers' movements. If the strobe is slowed down, the dancers move a greater distance between each flash that looks more like a series of individual poses than continuous motion. The vertical lines in Figure 4.1 below indicate when the voltage of

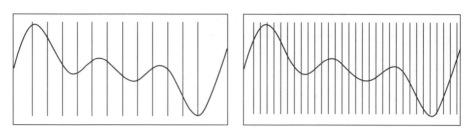

Figure 4.1 Identical signal sampled at two different sampling rates.

the input waveform will be sampled. The waveform depicted in the right panel is sampled more often than the sound depicted in the left panel.

The sampling rate in digital audio refers to the number of audio "snapshots," or samples, per second that have been taken for that signal. The higher the sampling rate, the better the digital characterization of the input signal. Sampling theory dictates that to uniquely represent any given input frequency, we need to collect at least two samples per cycle of that waveform. This means we have to use a sampling rate equal to at least double the highest frequency we wish to represent in our output signal. Conventional music CDs use a 44 100 Hz sampling rate, achieving a spectral bandwidth over 20 000 Hz that covers the full range of human hearing. Low sampling rates, on the order of 8 kHz and below, introduce large amounts of quantization error in time and throw away all high-frequency information above 4 kHz.

Another important factor in digital signal processing is the number of binary digits (or bits) used to represent each instantaneous measurement. This is called the resolution or word length. The greater the resolution, the more accurately each measurement is represented. A three-bit word has eight unique combinations of 1s and 0s. Given an input signal with a range of $+/-1$ V, a 3-bit is only able to resolve changes in the signal's voltage on the order of approximately 0.25 V. Figure 4.2 shows an input analog signal and it's 3-bit digital representation.

As you can see, there is some error introduced into the resulting digital signal from trying to represent a continuous analog signal with discrete, stepped digital data. The problem arises when the analog value being sampled falls between two digital values. When this happens, the analog value must be represented by the nearest digital value, resulting in a very slight error. The difference between the continuous analog waveform and the stair-stepped digital representation is known as quantization error. The quantization error introduced by this 3-bit sampling is shown below in Figure 4.3. This graph results from subtracting the digitized signal from the input analog signal.

A combined graph of the original input signal, the resulting digital signal and quantization error of a 3-bit sampling system is shown Figure 4.4.

A byte, or 8-bit word, has 256 unique combinations of 1s and 0s, and can represent the same 2 V range with greater precision. A 16-bit word can divide that same 2 V range into 65 536 equal steps. The process of dividing the continuous input voltage range of an analog signal into discrete digital steps is known as amplitude quantization. Similar to the sampling rate discussed above, the more steps you have, the better the digital characterization of the signal. Figure 4.5 shows the resulting quantization

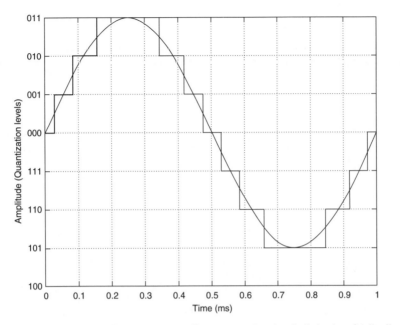

Figure 4.2 A single cycle of a sine wave after conversion to digital using 3-bits; the signal level is rounded to the nearest quantization level at each sample.

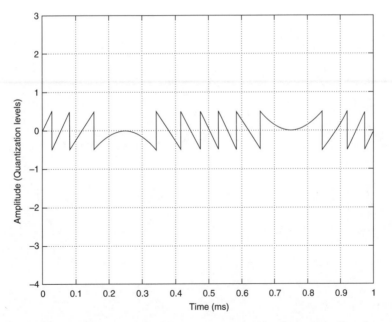

Figure 4.3 A graph of the quantization error generated by the conversion shown in Figure 4.2.

Audio Programming for Interactive Games

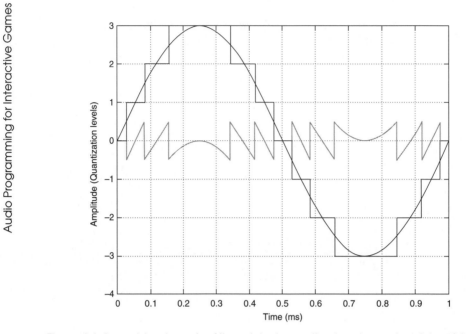

Figure 4.4 A combined graph of the original, quantized and error signals in a 3-bit system.

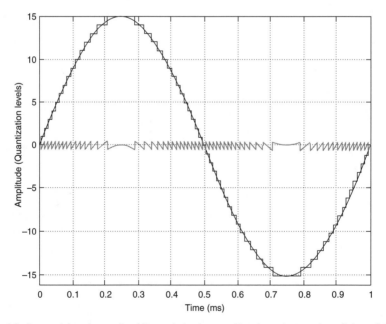

Figure 4.5 A combined graph of the original, quantized and error signals in a 5-bit system.

error in a 5-bit system. The error is significantly reduced, and the resulting digital signal more closely represents the original input analog signal.

Quantization error is one reason larger word lengths and higher sample rates sound better to our ears; the "steps" become finer, reducing quantization errors. Therefore, as a general rule, the greater the digital word size, the finer the digital representation of the input signal's amplitude, and the less quantization.

This doesn't mean all our problems go away with higher sampling rates and word sizes. From a sound quality perspective, it is important to recognize that the quantization error stays constant at $+/-0.5$ least-significant bit no matter what the input signal level. This means that the lower the signal, the more audible the error. As the sound decays from maximum to nothing, it uses fewer and fewer quantization levels and the perceived quality drops because the error becomes greater. This is significant because it is a distortion of the signal and is therefore directly related to the signal itself. Consequently the brain thinks that this is important material and pays attention to that distortion.

It's everywhere!

In the modern-day reality of fast desktop computers, CDs, DVDs, MP3 downloads and personal music players, it seems as if digital audio has always been an integral part of our daily lives. But in fact, it is a relatively recent occurrence. The first CDs didn't appear until the early 1980s, and it wasn't until the late 1990s that the widespread distribution and sharing of compressed DA, especially on the Internet, became possible and popular.

Storage space, the final frontier

The widespread use of digital audio in games is also a recent phenomenon. Early games did not use much, if any, digital audio because those computers and game consoles didn't have the throughput or storage capacity required of it. For example, CDs use a 44 100 Hz sampling rate and a 16-bit word size in stereo. Simple multiplication yields a data rate of 176 400 bytes per second or over 1.4 Mbits per second for this signal (44100 * 16 * 2)! To store just one minute of CD-quality audio data requires over 10 MB of storage. A 780 MB CD, therefore, can only hold approximately 74 min of sound. For a game that has to ship on a CD, including all the game code and visual resources, CD-quality audio takes up too much room.

In games, digital audio first appeared in hardware in the form of synthesizer chips. Since there were precious few resources to store previously sampled sounds, these devices generated samples on the fly using dedicated hardware. These samples were then converted back into an analog signal that we could amplify and hear.

It's speedy

Digital audio became more prevalent in games as computers became more powerful and storage more plentiful. Getting high-quality audio off the disk in a timely fashion still required a lot of CPU cycles, and storage prices remained high. For these reasons, low sampling rates (around 8 kHz) and 8-bit words were the norm. Audio compression technologies emerged to maximize the amount of data you could fit onto the distribution medium. Adaptive Delta Pulse Code Modulation (ADPCM) was one of the first audio compression technologies, reducing bit rates and storage requirements by a factor of four. It enjoys widespread deployment to this day due in no small part to the free distribution of one flavor of this compression scheme provided by the Interactive Multimedia Association (IMA-ADPCM).

By the late 1990s, it was possible to load some short sounds into memory before the game or level started, but any dynamic loading or swapping of those sounds during gameplay took more time and resources than the frame rate of the game could tolerate, so it generally did not happen. Hardware solutions, in the form of PC soundcards, or onboard sound chips in various game consoles, still ruled the roost. At best this hardware took some of the processing burden off the main CPU. At the very least, they were capable of generating musical tones and sound effects in real-time. Audio artists were forced to compose scores and sound effects for an underlying, often bad-sounding, audio chip using either Standard MIDI Files (SMFs), or a game platform's proprietary language and commands.

"Oh, woe is us!," came the cry. Mired in the muck of 2- and later 4-operator FM (Frequency Modulation) chips, countless hours were spent trying to tease out listenable tones from paltry sound hardware. This is whence the undeserved reputation of "bad MIDI music" comes. The initial use of MIDI in games was driven primarily by its compact and convenient way to make some noise on an aurally challenged game system. MIDI is a control language consisting of a stream of commands that can, among other things, direct a receiving device to change instruments, turn notes on and off, and set a sound's volume and pan positions. It makes no sound in and of itself, and therefore cannot "sound bad." This is a fallacy! SMFs are widely used because they are orders of magnitude smaller than their digital audio counterparts, at any resolution. They consist entirely of music command sequences; they are digital musical scores. This makes MIDI an attractive and practical solution to the problems of digital audio described above. Today, MIDI is often coupled with high-quality digital audio inside a software synthesizer for games. We'll take a closer look in the Chapters 5 and 7 and learn how to do this. But let us not dwell on the past. Let us instead learn the lessons the past has to teach, forgive and move on.

Don't look back

In modern computer games, DA is here to stay. From sound effects to dialog to massive background tracks and full orchestral scores, DA is used everywhere.

The capabilities of game platforms, both desktop machines and dedicated consoles, have risen over the past several years to where 16-bit, stereo files sampled at 22.05 kHz are an ordinary occurrence. This rise in the quality and ubiquity of digital audio is due to several factors. Storage capacity has greatly increased per unit area, and the price of storage has dropped precipitously. Modern audio perceptual compression technologies commonly achieve 10:1 to 20:1 compression ratios, and beyond (!), with little reduction in perceived sound quality, at least for most listeners. We can squeeze more into a single unit area, and have more space with which to work. Furthermore, computers have gotten blindingly fast and can easily handle the data rates of high-quality digital audio on the main CPU. The sophistication and complexity of dedicated hardware has also greatly improved.

Get in line, there

In the majority of nongame applications, DA is played linearly. That is, you hit the start button, and out it comes. Each time you play a piece of DA, you typically listen to it from start to finish. If it takes some small amount of time to cue up the song or start playing the track, it's no big deal. As long as the music doesn't stop or pop or click or skip once it's begun, we're happy.

In games, however, digital audio is consumed in stages. Background tracks utilize multiple, overlaid sounds to create a dynamic soundscape. Titles now use as many as 40 000 lines of dialog, and sound effects fly fast and furiously in, out and around your head. Games turn the notion of linear playback completely on its head. In an interactive game, each player creates his or her own audio experience as the game is played. To a very large degree, we can never know what sounds will follow one another. OK, maybe 'never' is too strong of a word. But it's darn close.

We're goin' in

As was mentioned at the outset of this chapter, the goal of the Soundtrack Manager is to support digital audio playback in an interactive and musical fashion. This requires us to not only have access to the digital audio samples themselves, but to a number of additional descriptive and control parameters to manage that playback. These parameters, along with the information we need to get at the raw digital audio samples, are bundled by the Soundtrack Manager in what is called a SAMPLE structure. All of this information shows how the Soundtrack Manager represents and stores digital audio data. The specific members of this structure will change across different platforms to accommodate platform-specific output needs. When this happens, all affected files are kept in a separate folder for that particular version of the software. But the Soundtrack Manager API uses SAMPLEs, meaning most of the code remains unchanged from one platform to the other. The SAMPLE structure for the DSMIX version of the Soundtrack Manger is shown below (found in \Code\Sound\Win9X\DSMix\Sample.h):

```
typedef struct _sample_t {
  char          *fileName;
  UINT16        bitWidth;
  UINT16        nChannels;
  UINT32        length;
  UINT32        size;
  void          *compressionInfo;
  UINT16        volume;
  INT16         pan;
  UINT32        freq;
  UINT8         *data;
  BOOLEAN       isStreaming;
  BOOLEAN       isPlaying;
  BOOLEAN       isPaused;
  BOOLEAN       isLooped;
  streamdata    *streamData;
  FADEINFO      *fadeInfo;
  loop_t        *loop;
  SAMPLEMIDIINFO MIDI;
  struct _sample_t  *next;
  struct _sample_t  *origSample;
  struct _sample_t  *playNext;
  void (*DonePlayingCB)(struct _sample_t *sample);
  FILEHANDLE hFile;
  UINT32        fileDataOffset;
  UINT32        curFillPosition;
  UINT32        curPlayPosition;
  BOOLEAN       (*ResetFile)(void *sample,
                UINT32 fileDataOffset, UINT32 dataOffset);
  BOOLEAN       (*ReadFile)(void *sample, UINT32 length,
                UINT8* data, UINT32* uChkErr);
  LPDIRECTSOUNDBUFFER lpDSB;
} sample_t, SAMPLE;
```

As you can see, this structure is quite complex. It not only contains the DA data itself, in the form of a file name or memory pointer, but a lot of information about and properties of that data: its sampling rate, bit depth, number of channels and length (in bytes). It also holds any information we may need to decompress the data, set its volume or pan position and whether it is being streamed from disk. Performance information is also maintained: whether the sample is playing, paused, looped or fading and the information necessary to use it as a wavetable instrument for MIDI playback (see Chapter 7). It keeps track of where we are in the file, and our current fill and play positions to feed our streaming buffers. There are also several callback function pointers we can assign at runtime to signal when the SAMPLE is done playing, read data into the SAMPLE structure, or reset the SAMPLE to its starting position.

We are currently examining the DSMIX version of the Soundtrack Manager in which we use DirectSound to mix our SAMPLEs together. Therefore, the final element of our SAMPLE structure is a pointer to a Windows DirectSound buffer. In another version of the software, we do the mixing ourselves using floating-point assembly instructions (see Chapter 8). The SAMPLE structure in that project does not contain any DirectSound references, but the names of the files (Sample.c and Sample.h) are the same. This is another way we support the re-configuration of the Soundtrack Manager for specific platforms or to accommodate specific design changes. The file names in our Visual C++ projects remain the same, but their contents are different and so are placed in different folders to keep them separate.

What's inside a SAMPLE?

The SAMPLE Soundtrack Manager data type may reference any size chunk of DA data. It may be a short sound effect, a lengthy spoken narrative, a recording of single instrument note, an entire orchestral cut-scene, and anything in-between. Considering what we know about digital audio sampling and size requirements, if the SAMPLE contains a short sound, we can probably load it all into memory at once and play it back from there. However, if the SAMPLE is long, we may not have enough memory or time to load all the data all into memory before we have to play it back. In these situations, we can choose to play it back from the disk by grabbing and playing small chunks of the file one right after the other. This process is called "streaming" and is used all the time in games for long digital audio clips. Within the SAMPLE structure, streamData a pointer to a streamdata structure that holds all the information necessary to stream this SAMPLE:

```
typedef struct _streamdata {
  UINT32    bufferSize;
  UINT32    bufferSegSize;
  UINT32    segOffset[NUM_BUFFER_SEGMENTS];
  UINT32    nextWriteOffset;
  UINT32    jumpToOffset;
  UINT32    lastUpdatePos;
  BOOLEAN   inRelease;
  BOOLEAN   foundEnd;
  UINT32    lengthToStop;
  BOOLEAN   isReset;
} streamdata;
```

Here we keep track of the total size of the memory buffer we'll use to stream the audio data. Since we can't read and fill the same memory locations simultaneously, we cut up that memory buffer into at least 2 segments (NUM_BUFFER_SEGMENTS) so we can play back one segment while filling the other with the next audio chunk, a process known as "double-buffering." We also keep track of several offset positions necessary for filling our SAMPLE's DirectSound buffer, and store some flags about where we are in the data itself and what data we should retrieve next time around.

I'm feelin' loopy

Sounds can be looped in the Soundtrack Manager. The loop element in the SAMPLE structure is a pointer to a SAMPLELOOP structure that specifies the beginning and ending sample numbers between which we want to loop. These loop points are not authored into the file, but can be set by a programmer (and later, the audio artist – see Chapter 10). The SAMPLE may also be used as a note in a MIDI software synthesizer. The SAMPLEMIDIINFO data structure keeps track of the information necessary to accommodate this use of a SAMPLE. (We'll explore this in more detail in Chapter 7.)

Again, all of the information presented above shows how the Soundtrack Manager represents and stores digital audio data. But how does it get played? To answer this, we have to shift gears for a bit and talk about the underlying mechanisms that make the Soundtrack Manager tick.

Making it work

Critical sections

After linking the Soundtrack Manager library with their game code, the first thing the game programmer has to do is call the SMInit function in the file SM.c (located in the \Code\Sound folder). Here we initialize a few variables, including one to count how many times we've entered a critical section in our code. This is a set of statements that can have only one process executing it at a time. In a multi-threaded operating system, such as Windows, an application can have multiple threads, or independent paths of execution, running simultaneously. This is useful if the application needs to manage multiple activities, as in a game. Individual threads can be set up to handle different tasks such as updating the display screen, accessing disk files and getting data from a communications port. The bad news is that we can never be sure what thread will be executing at any one time, and several different threads can share the same data. Consequently, there are times when one thread should not, under any circumstances, be interrupted by another. For example, if thread 1 is allocating or de-allocating memory, or re-ordering a linked list that is also used by thread 2, those operations must be completed before thread 2 can execute. If we allow thread 2 to run prematurely, we will likely experience some very unpredictable behavior, or worse, end up in the nowhereland of the blue screen. Therefore, we use a "critical section" object for mutual-exclusion synchronization to protect us from our own worst enemy – ourselves. One other cool thing about critical sections (under Windows) is that once a thread has taken ownership of a critical section, it can make additional calls to claim that critical section without blocking its own execution. This prevents a thread from deadlocking itself while waiting for a critical section that it already owns. We also keep a count of the number of times we've entered our critical section object, just for fun.

Playing nice

Next we initialize the global variable `active` to `FALSE`. On the PC, a user can have several different programs running simultaneously, and switch back and forth and among them at their leisure. Each program, including a game, must be polite to one another when they gain focus (become active) or lose focus (become inactive). System resources should be relinquished, sounds and video should stop playing; that is, all programs should get out of the way of each other as directed by the computer user. No matter how cool your game, you have to stop running when the player switches back to their spreadsheet application because their boss is heading their way. We'll see how we accommodate this change of focus in a little bit. But at this point in our initialization, we're not ready to run, and indicate the same.

Memory and loudness

Next we initialize our memory management system. Memory management is another place where different platforms have different ways of doing things. We accommodate these differences by defining platform-specific macros for memory allocation and deallocation (in `\Code\Sound\Win9x\SMPlatfm.h`). Called `APPMALLOC`, `APPFREE` and `APPMEMSET`, these macros can be set to any platform or application-supplied methods for handling memory. In the `InitMemMgr` function (found in `\Code\Sound\SMMemMgr.c`), we can track the current number of allocations, their associated pointers and total size by defining the symbol `__SLIB_MEMORYLEAK_DEBUG__`. If the symbol is not defined, that debugging overhead goes away.

In `InitVolumes` (found in `\Code\Sound\Volume.c`), we set up two `MASTER-VOLUME` structures, `MusicMasterVolumePtr` and `SFXMasterVolumePtr`. These allow us to save and control the volume, pan and fade of all our music and sound effects as a whole with one global structure. Having this one structure means we don't have to keep track of and individually control all those sounding elements in a specific group:

```
typedef struct _mastervolume {
  UINT16      volume;
  INT16       pan;
  PROCLIST    *procList;
  FADEINFO    fade;
} MASTERVOLUME;
```

The `procList` member points to a list of dynamically allocated function pointers we wrap inside a `PROCLIST`. We add procs to this list by calling `MasterVolumeProc-ListAdd` (in `\Code\Sound\Volume.c`) to adjust those resources that are part of a particular volume group. For instance, we add the `MIDIFileSetMusicMaster-Volume` function to the `MusicMasterVolumePtr` in `InitMIDIFiles`. We'll talk about the fades and the `FADEINFO` data type a little later on in this chapter.

Audio I/O devices

All the devices we've enabled for digital audio and MIDI playback are initialized next in `InitDevices`. MIDI handling routines are set up in `InitMIDIHandlers`, and a MIDI voice table is initialized by calling `InitVoiceOwners` (all of this code can be found in `\Code\Sound\Win9x\Devices.c`). These `VOICEOWNER` structures will be used to track which sounds are being played by what MIDI notes in our software synthesizer (described in Chapter 7).

Next we claim the DirectSound devices we'll be using in this version of our library by calling `PCMDeviceInit`, `MIDIDeviceInit` and `MIDIInDeviceInit` (in `\Code\Sound\Win9x\DSMix\PCMDev.c` and `\Code\Sound\Win9x\MIDIDev.c`, respectively). Each of these routines will be replaced in different versions of the library to accommodate specific platforms or design changes. The remaining routines that get called in these functions have to do with our MIDI synthesizer, so we'll defer any discussion of these until Chapter 7.

Modular planning

Coming back to `SM.c`, we initialize each of the modules defined in `modules.h`, (described in the previous chapter), and create four linked lists of function pointers to control and manipulate these various modules throughout the run of the game. A closer examination of the `ModuleProcs` array (just above the `SMInit` code itself) reveals the specific functions we will use to initialize, uninitialize, activate, deactivate and update each of our defined modules. Not all the modules need to be activated or deactivated, and therefore have NULL pointers in those locations. Upon the successful initialization of each module, we traverse the `ModuleProcs` array and copy the stored function pointer, if there is one, into the appropriate linked list for that activity, like so:

```
i = 0;
while(moduleInitProc = ModuleProcs[i][0]) {
  // if no init proc, end of the list has been reached
  if(status && moduleInitProc()) { // execute init proc
    if(moduleUninitProc = ModuleProcs[i][1]) {
      ExitProcList = ProcListAddHead(ExitProcList,
                              moduleUninitProc);
    }
    if(moduleActivateProc = ModuleProcs[i][2]) {
      ActivateProcList = ProcListAddTail(
        ActivateProcList, (void*)moduleActivateProc);
    }
    if(moduleDeactivateProc = ModuleProcs[i][3]) {
      DeactivateProcList = ProcListAddHead(
      DeactivateProcList, (void*)moduleDeactivateProc);
    }
    if(moduleUpdateProc = ModuleProcs[i][4]) {
      UpdateProcList = ProcListAddHead(UpdateProcList,
                              moduleUpdateProc);
    }
  } else {
    status = FALSE;
  }
  i++;
}
```

The first of these procedure lists is the ExitProcList. Each of the modules compiled into the library specify their own cleanup procedure to release and free any resources and memory they may be using. Every module that is compiled into the library has a corresponding exit function that will be called when the game is shut down.

The next two procedure lists that get created are the ActivateProcList and DeactivateProcList. For audio, when the game loses focus, we have to immediately stop all current playback to make room for whatever application the user is switching to. When the game regains focus, we should resume all audio playback at the same location(s) we were at when we lost focus. We handle these situations cleanly via the functions in these two PROCLIST lists.

The last procedure list we build is the UpdateProcList. As we'll see below, the Soundtrack Manager sets up a high-resolution timer to generate periodic interrupts and calls the SoundManager function to do its work. For each module included in the library, a corresponding update function is inserted into this list and called during those timer interrupts. It is in these update functions that we handle the bulk of the Soundtrack Manager's affairs.

On your mark, get set . . .

Once we have successfully constructed the procedure lists, it is time to start the engine running. The `InitPlatform` function is called to initialize the platform on which we're running (in `\Code\Sound\Win9x\SMPlatfm.c`):

```
if(status && InitPlatform()) {
  SMInfo.frame = 0L;
  SMInfo.frameRate = PlatformGetFrameRate();
  return(TRUE);
}
```

This routine will be different for each platform on which we find ourselves, and will again be stored in a different folder. As with other structures and routines detailed in this chapter, this is how we support different platforms within our Soundtrack Manager. Notice, too, this is where the critical section object itself is initialized. Below is the code for the PC (residing in `\Code\Sound\Win9x\SMPlatfm.c`):

```
BOOLEAN InitPlatform(void)
{
  SMOptions.platformInfo.platformFunc = NULL;
  InitializeCriticalSection(
       &(SMOptions.platformInfo.criticalSection) );
  if(InitPlatformTimer()) {
    PlatformInitialized = TRUE;
    return(TRUE);
  }
  DeleteCriticalSection(
       &(SMOptions.platformInfo.criticalSection));
  return(FALSE);
}
```

The final piece of this part of our Soundtrack Manager puzzle is the `Init-PlatformTimer` function (also in `\Code\Sound\Win9x\SMPlatfm.c`). We must regularly update all our sound resources to keep the audio flowing, and set up a 125 Hz timer to do so. This is where we set up the high-resolution timer that will drive the operation of the Soundtrack Manager from this point forward.

```
// Set up main event timer
if(timeGetDevCaps(&tc, sizeof(TIMECAPS)) ==
    TIMERR_NOERROR) {
// Set timer accuracy to 1mS
pInfo->wTimerRes = MIN(MAX(tc.wPeriodMin, 1),
    tc.wPeriodMax);
// Start the timer.
if(timeBeginPeriod(pInfo->wTimerRes) !=
    TIMERR_NOERROR) return(FALSE);
// Set up timer's callback function.
```

```
if(pInfo->wTimerID=timeSetEvent(SM_TIMER_PERIOD,
  pInfo->wTimerRes, (LPTIMECALLBACK)PlatformTimeProc,
  (DWORD)(SoundManager), TIME_PERIODIC)) {
  pInfo->timerCount = 0;
  pInfo->frameRate = 1000 * 4096 / SM_TIMER_PERIOD;
  pInfo->bTimerInstalled = TRUE;
  return(TRUE);
} else {
  //Error setting up Windows timer callback
}
...
```

In our sound engine, we set up a periodic timer to call `PlatformTimeProc` every 8 ms, equivalent to an update rate of 125 Hz. This function calls `SoundManager` which in turn executes all the functions in `UpdateProcList`, thereby updating all our defined modules in one easy call.

GO!, er, oops . . .

That's essentially all there is to `SMInit`, except if something goes wrong (gasp!). Should we encounter an error anywhere along the line, we simply back out the way we came in, leaving no trace of our presence, no soda cans in the bushes:

```
        ProcListExecute(ExitProcList);
        ProcListFree(ExitProcList);
        ProcListFree(UpdateProcList);
        ProcListFree(ActivateProcList);
        ProcListFree(DeactivateProcList);
        UninitDevices();
          }
        UninitVolumes();
      }
    UninitMemMgr();
    }
  return(FALSE);
}
```

Play me

Let's return to how we actually play back DA data. In the Soundtrack Manager, all DA is stored inside a `SAMPLE` structure for playback. `SAMPLE` structures are created in one of several ways: DA is either read from a file, loaded into RAM (in part or in whole), or copied from an existing `SAMPLE`. We keep a linked list of all of our active

SAMPLE structures so we can easily tend to all current digital audio resources during the periodic update interval. We also keep a list of those SAMPLEs we have to restart when our game gets deactivated. These two lists get initialized in the InitSamples routine (in \Code\Sound\Win9x\DSMix\Sample.c):

```
BOOLEAN InitSamples(void)
{
  Samples = NULL;
  SamplesToRestart = NULL;
  return(TRUE);
}
```

As you would expect, the contents of Samples changes dynamically over the course of the game as different SAMPLEs are loaded, played and discarded.

Here's how to load and play a WAV file using this library:

```
SAMPLE *sample = NULL;
if(sample = GetSampleFromFile("MyFile.wav", FALSE,
    FALSE, NULL)) {
  PlaySample(sample);
} else {
  //Error loading WAV file
}
```

That's it, just a few lines of code! To see what's really going on behind the scenes in this rosy scenario, please read on as we pull back the curtain in the next few paragraphs.

Pulling back the curtain

The GetSampleFromFile function (located in \Code\Sound\Win9x\DSMix\ Sample.c) takes four arguments. The first and most obvious one is the file name itself. We show a common WAV file in our example, but it could just as easily have been an AIF file, an ADPCM file wrapped up in a WAV, or a file containing raw digital audio samples with no other distinguishing characteristics. The type of digital audio formats supported by the library are defined in modules.h (discussed in Chapter 3), and can change and grow as necessary to fit your needs.

The next two arguments to this function are simply flags instructing the Soundtrack Manager whether to loop the sound or to stream it or not. In this example, we've chosen to do neither.

The final argument to GetSampleFromFile specifies a callback function to call when the sound is done playing. This feature is useful because you can start playing a sound, and not worry about it again until it's done. At which point, the "done" function will be called. While we don't specify any such function in our simple example above, this callback can be used to coordinate the audio presentation in a game in a musically significant way. You don't have to guess when the current sound is finished to start or stop new sounds. You can cue up new music, or take some other appropriate action based on what's currently happening in the game. You know when the sound is done because the Soundtrack Manager tells you. A very powerful use of this signaling mechanism is to specify a callback function in the game code itself. In this way, the sound can drive the game's behavior, not the other way around. For instance, at the end of a level, the game can wait until the final music is done playing before displaying the interstitial graphics for the next level. While the audio programmer typically sets up this callback, the game programmer can also do so since the `GetSampleFromFile` function is public.

The `GetSampleFromFile` function takes all this information, allocates space for a new `SAMPLE` structure, and tries to determine what kind of digital audio file it's been given. Based on the defined `SAMPLE` submodules, we open and parse the specific type of digital audio file we've been given. In our `WAV` example, we call `GetSample-FromWAV` and fill out the new `SAMPLE` structure itself. We set the `fileName` field, set it's volume to `MAX_SM_VOLUME`, set the `loop` and `isStreaming` flags appropriately, and set the `DonePlayingCB` member. Two important function pointers are also set to rewind the file to the beginning of the data, `WAVResetFile`, and to handle all reading from the file from this point forward, `WAVReadFile`. In the case where the file is to be looped or streamed, we also allocate a `SAMPLELOOP` and `streamdata` structure inside the new `SAMPLE`, respectively.

Once a `SAMPLE` has been loaded, we can clone it using the `GetSampleFrom-Sample` routine. All the information from the first `SAMPLE` is copied into a second one, and it gets inserted into the global linked list of `SAMPLE`s we're currently tracking. We can make multiple copies of a sound, such as a gunshot, using this facility without having to open the original file over and over again.

In this `DSMIX` version of our Soundtrack Manager, we let DirectSound mix all our `SAMPLE`s together. This requires us to create a DirectSound secondary buffer for each `SAMPLE`, shown in the `SampleCreateBuffer` function (also in `\Code\Sound\Win9x\DSMix\Sample.c`). You can find many tutorials on how to use DirectSound, so we won't go into that here.

Finally, we fill the sound buffers with data, either in whole or in part (for streaming purposes), and add the `SAMPLE` to our global linked list of active `SAMPLE`s. Note that this happens inside a critical section, because we don't want anyone to use the list whilst we're modifying it:

```
if(SampleResetBuffer(sample)) {
  //Prepend sample to global linked list of SAMPLEs
  PlatformEnterCriticalSection();
  sample->next = Samples;
  Samples = sample;
  PlatformExitCriticalSection();
  if(!stream) {
    SLib_FileClose(*hFile);
    *hFile = NULL;
  }
  return(sample);
}
```

The last few statements close the input file if we're not streaming, as we've loaded all the data from the file into memory. All is well, and we return our newly created and filled SAMPLE to the caller for further instructions.

In our current example, begun now several pages back, those instructions are to call PlaySample. The entirety of this code is inside a critical section, as it changes the state of a number of variables, fills a SAMPLE's buffers, if necessary, and starts the SAMPLE playing.

Alright! We did it!

Keep me posted

You will recall from the discussion above that PlatformTimeProc gets called each time the periodic timer expires, which in turn calls our workhorse routine, SoundManager. Here we update all our master volumes, then call ProcList-Execute to update all our modules. One of the function pointers in this update list is UpdateSamples. Inside this function, we traverse the list of active SAMPLEs. If they're not streaming, we don't do anything. But if they are streaming, we adjust their individual volumes if necessary, fill their buffers with the next chunk of digital audio data or stop them if they're done playing. This last action issues a stop command to the DirectSound buffer, sets the SAMPLE's isPlaying flag to FALSE, and calls the DonePlayingCB function, if there is one. Note that the callback function takes a pointer to the SAMPLE structure being stopped. This function is often charged with handling all SAMPLE callbacks, and can identify the specific SAMPLE that has stopped via this argument. This provides another point where we can control the overall flow of the sound in a game. If the callback comes to a Soundtrack Manager function, we may want to replay that SAMPLE or start any number of other SAMPLEs, depending on the game's progress at that time. If the callback goes to the game code itself, the game could synchronize the graphics of the game to what the music is doing. The Soundtrack Manager callback routine can call into the game code as well. The possibilities are endless, and provide the creative opportunity to tightly coordinate the aural and visual presentations of the game.

Get out of the way

The final two pieces of this SAMPLE puzzle come into play when a game loses focus. On platforms such as the PC, the user can have many different applications running simultaneously, and can switch back and forth and among them at will. Following good programming etiquette, a game, or any application, should release any shared system resources it may have claimed before it is deactivated. It must also reclaim those resources when the game regains focus. In the Soundtrack

Manager, just before the game loses focus, the game should call `SMDeactivate` (in `\Code\Sound\SM.c`):

```
BOOLEAN SMDeactivate(void)
{
  if(active) {
    if(PlatformDeactivate()) {
      if(DevicesDeactivate()) {
        if(ProcListExecuteB(DeactivateProcList)) {
          active = FALSE;
          return(TRUE);
        }
      }
    }
  } else {
    return(TRUE);
  }
  return(FALSE);
}
```

Any deactivation operations that may need to be performed for the underlying platform happens here. `SAMPLE`s use the `__SLIB_DEVICE_PCM__` device, which gets deactivated in `PCMDeviceDeactivate` (in `\Code\Sound\Win9x\DSMix\PCMDev.c`). In the `DSMIX` version of the Soundtrack Manager, this means we also release the underlying DirectSound object.

Back in `SamplesDeactivate`, our list of active `SAMPLE`s is copied into another global linked list of `SAMPLERESTARTLISTENTRY`s, cleverly named `Samples ToRestart`. Here we save each `SAMPLE` structure's state along with its volume and pan information. We call `PauseSample` on each `SAMPLE`, set `isPaused` to `TRUE`, and save the current play position for sample-accurate resumption when the game is reactivated. We release the DirectSound sample buffer, and we're out.

Jump right in

To activate the Soundtrack Manager when the host application is brought back into focus, we do the same operations in reverse when the game calls `SMActivate`:

```
UINT32 SMActivate(void) {
  UINT32 code = SM_ERROR;
  if(!active) {
    if(ProcListExecuteB(ActivateProcList)) {
      if((code=DevicesActivate()) == DEVICES_OK) {
```

```
        if(PlatformActivate()) {
          active = TRUE;
          return(SM_OK);
        }
      }
    }
  } else {
    return(SM_OK);
  }
  return(code);
}
```

`SMActivate` calls `ProcListExecuteB` to call each of the functions in the `ActivateProcList`, which includes the function `SamplesActivate`. We reallocate and refill all the DirectSound buffers, both streaming and one-shots, restart all those `SAMPLE`s in our `SamplesToRestart` list, and we're off once again.

Some other neat things

We can stop or pause a playing `SAMPLE` at any time by calling `StopSample` or `PauseSample`, respectively. While both of these functions immediately stop the sound, `PauseSample` saves the `SAMPLE`'s current play position. This allows the `SAMPLE` to resume playback from the point it was paused on any subsequent `PlaySample` call. `StopSample`, on the other hand, resets the play position to the beginning of the sound.

It is also possible within our Soundtrack Manager to segue from one playing `SAMPLE` to another with two simple calls. `SampleAppendSample` sets the `playNext` field of the current `SAMPLE` struct to point to the second sound. Calling `SegueSample` on that `SAMPLE` will immediately start the second sound and stop the first one. If the `playNext` field is empty, a call using this `SAMPLE` to `SegueSample` will simply stop the sound.

In a game, we have to be able to dynamically change the volumes of the sounds. This ability not only allows us to compensate for level differences among the sounds themselves, but lets us set and change the audio mix during and in response to gameplay. For example, we may need to "duck" the background music while we play a voiceover, or a sound may start softly and grow in intensity as it moves closer to you. To accommodate this dynamic behavior, our Soundtrack Manager API presents three functions to get and set the gain of an individual `SAMPLE`. The first two are called `SampleGetVolume` and `SampleSetVolume`. They both take a `SAMPLE` pointer and a volume and pan parameter, and either get or set a sound's volume and pan position inside the mix.

The `SampleSetFade` function fades a sound in or out to some target volume and pan position over a specified period of time. In it's argument list, it takes a pointer to a `SAMPLE`, the fade's target volume and pan, the number of milliseconds over which to perform the fade, and a flag to stop the sound playing at the end of the fade or not. This functionality is made possible by our underlying timer and update architecture. If you recall, the `UpdateSamples` function is called every 8 ms. If the `fadeInfo` field of the current `SAMPLE` is defined, we call `GetFadeVolume` to calculate the fade's current volume and pan values and decide whether those parameters need to be updated in this frame or not. If so, we call `SampleSetVolume` to make the necessary changes, and check if we're done fading. At the end of the fade, we stop the `SAMPLE` if the `stopWhenDone` field of the `FADEINFO` structure is set. Otherwise, the sound continues to play at the target volume and pan position. Once the fade is complete, the `fadeInfo` structure is cleared:

```
if(GetFadeVolume(fade, &volume, &pan, &update,
   &doneFading)) {
 if(update) {
  SampleSetVolume(sample, volume, pan);
   if(doneFading) {
    if(fade->stopWhenDone) {
     StopSample(sample);
    }
    ClearFadeInfo(fade);
   }
  }
 }
}
```

To loop a `SAMPLE`, we call the `SampleSetLoop` function. This function takes a `SAMPLE` pointer, a flag to loop the `SAMPLE` or not, and the beginning and end points of the loop itself. In this version of the library, however, DirectSound cannot accommodate this kind of controls. Therefore, both steaming and non-streaming sounds will loop from start to finish, regardless of the points specified to this function. `SampleGetLoop` simply returns the `SAMPLE`'s loop status (yes or no), and the sample points of the loop.

There is an additional parameter to both these calls that we have not discussed, and that is the `pingPong` argument. We use this parameter to modify the nature of the looping itself. Normally, sounds are looped from the first point to the last, over and over again. Setting `pingPong` to `TRUE` causes the sound to loop forward and backward between the two points: first to last, last to first, first to last and so on. Unfortunately, this behavior is also not supported in DirectSound itself, so setting this flag will have no effect on the sound's looping behavior in this `DSMIX` version of our library. In Chapter 8, however, we bypass most of DirectSound by writing a custom mixer, and show how to accommodate this and a number of other extended `SAMPLE` behaviors.

When we are all done with a `SAMPLE`, we call `FreeSample` to deallocate all the memory associated with that object, remove it from the global linked list of `SAMPLE`s, and free the `SAMPLE` data structure itself. All this is performed within a critical section to ensure all our housekeeping is finished before it's used or referenced by another thread.

There are a handful of functions remaining in our SAMPLE API, and most of these are informational and self-explanatory in nature. The first of these are SampleGetFreq and SampleSetFreq that get and set the sampling rate of the specified SAMPLE. The latter function changes the pitch of the SAMPLE, and comes with a rather big processor hit from DirectSound. SampleGetLength returns the size, in bytes, of the encapsulated digital audio data, while SampleGetPlayStatus returns the playing and paused status of the specified sound.

SampleSetPosition allows you to jump immediately to a specific point within your SAMPLE. For instance, long sounds are usually streamed, and multiple sounds can be encased within a single SAMPLE. You could set up one section of the sound to loop, and then call this function to jump to another sound later on in the data when a particular event happens in the game; for example, when the character opens a door and enters another room. SampleGetPosition returns the current playback position within a SAMPLE.

Summary

In this chapter, we have presented and discussed the Soundtrack Manager's public DA interface. We have shown how DA is loaded, played and manipulated via these functions, and described the inner workings of the library's interrupt-driven architecture. We have introduced a number of musically significant behaviors, and introduce how to support our goal of platform independence when using DA assets.

5 Interfacing with MIDI – Interpreter, please!

This chapter details the implementation of another game audio asset – MIDI (the Musical Instrument Digital Interface). The high-level musical services offered via MIDI are discussed, and a robust interpreter for both MIDI input and output streams is presented. Platform-specific operations are again compartmentalized for easy replacing when moving to a different platform.

More on devices

As we discussed in the Chapter 4, the Soundtrack Manager plays a game's audio content through some platform object we call a `device`. For example, the output device for digital audio on the PC is the soundcard, accessed through `DirectSound`. On the Macintosh platform, the digital audio output device is often the built-in `SoundManager`. The digital audio output device on the PlayStation2 is the embedded `Sound Processing Unit (SPU)`. Inside our library, we keep an array of devices, called `Devices`, to store all the platform-specific information we need to support digital audio, MIDI input and MIDI output.

The Soundtrack Manager's `device` structure is defined as:

```
typedef struct {
  UINT type;
  BOOLEAN active;
  union {
    MIDIdevice MIDI;
    MIDIindevice MIDIIn;
    PCMdevice PCM;
  } dev;
  struct _MIDIhandlers *MIDIHandlers;
} device;
```

The first two members identify the `type` of device, digital audio or MIDI, and if it is `active`. (Please see Chapter 4 for more thorough discussion of what it means for a device to be active or not). A union of our known device types follows next in dev. MIDIdevice, MIDIindevice and PCMdevice contain the platform-specific information and data structures necessary for each Soundtrack Manager device. A set of MIDI handler functions is associated with each device in MIDIHandlers.

The device structure affords another place for us to customize this library for the specific, underlying platform. On the PC, the PCMdevice contains pointers to the DirectSound object, the DirectSound primary buffer, a linked list of secondary buffer pointers, and two boolean variables to indicate DirectSound's initialization and active status. (For more information on how DirectSound works and what these elements are, please refer to the DirectSound documentation in the DirectX SDK from Microsoft). On another platform, we may only need the initialization and active flags. The point is, the device structure itself does not have to change, only the referenced PCMdevice definition. Make no mistake – building this library for a different platform does require additional work. In this particular case, we have to write new Devices.h and Devices.c files for every new platform we have to support. We place them in a different folder, and include them in a new project. But the names of the files, structures, functions and argument lists do not change, only their internal, platform-specific contents. All necessary modifications are localized to and under the control of the sound department. The big win is that from the game's or other multimedia application's perspective, the Soundtrack Manager looks the same on whatever platform we find ourselves.

MIDI services

Depending on the devices we've defined in modules.h (described in Chapter 2), we initialize the MIDI services we're going to use in our game by calling the Init-MIDIHandlers routine from within InitDevices (in \Code\Sound\Win9x\ Devices.c). The Soundtrack Manager makes liberal use of function pointers. This is so we can configure what function actually gets called at runtime instead of hard-coding them beforehand. This is another way we support different platforms through the same interface. With that in mind, the MIDIHANDLERS structure of a device is a group of function pointers wherein we specify the function names, argument lists and return types of the MIDI messages we want to support:

```
typedef struct _MIDIhandlers {
  BOOLEAN (*NoteOff)(VOICEOWNER *voiceOwner);
  VOICEOWNER *(*NoteOn)(UINT8 ch, UINT8 note, UINT8 vel,
                        INT8 pan);
  BOOLEAN (*PolyPressure)(UINT8 ch, UINT8 note,
                          UINT8 value);
  BOOLEAN (*Control)(UINT8 ch, UINT8 ctrl, UINT8 value);
  BOOLEAN (*Program)(UINT8 ch, UINT8 program);
  BOOLEAN (*AfterTouch)(UINT8 ch, UINT8 value);
```

```
    BOOLEAN (*PitchBend)(VOICEOWNER *voiceOwner,
                          UINT8 vhi, UINT8 vlo);
    BOOLEAN (*SetNoteVelocity)(VOICEOWNER *voiceOwner,
                          UINT8 vel, INT8 pan);
  } MIDIHANDLERS;
```

InitMIDIHandlers simply calls MIDIDeviceMIDIHandlersInit and PCM-
DeviceMIDIHandlersInit to set these function pointers to the real functions
they will call in this version of our library (in \Code\Sound\Win9x\MIDIDev.c and
\Code\Sound\Win9x\DSMix\PCMDev.c, respectively). For example:

```
  void MIDIDeviceMIDIHandlersInit(void) {
    MIDIDevMIDIHandlers.NoteOff = MIDIDeviceNoteOff;
    MIDIDevMIDIHandlers.NoteOn = MIDIDeviceNoteOn;
    MIDIDevMIDIHandlers.SetNoteVelocity =
                            MIDIDeviceSetNoteVelocity;
    MIDIDevMIDIHandlers.PolyPressure =
                            MIDIDevicePolyPressure;
    MIDIDevMIDIHandlers.Control = MIDIDeviceControl;
    MIDIDevMIDIHandlers.Program = MIDIDeviceProgram;
    MIDIDevMIDIHandlers.AfterTouch = MIDIDeviceAfterTouch;
    MIDIDevMIDIHandlers.PitchBend = MIDIDevicePitchBend;
    return;
  }
```

I hear voices

InitDevices next calls InitVoiceOwners to initialize a pool of VOICEOWNER
structures. These are used to track which sounds are being played by what MIDI
notes in our software synthesizer. (Further details on VOICEOWNERs and how they
are used can be found in Chapter 7 when we build a software wavetable synthesizer.)

Again, depending on the devices defined in modules.h, we claim the DirectSound
devices we'll be using in this PC version of our library by calling PCMDeviceInit,
MIDIDeviceInit and MIDIInDeviceInit. We set the device type of each device
in our global Devices array, and assign the MIDIHandlers field to the address of
the specific MIDIHANDLERS we wish to use for each device:

```
  #ifdef __SLIB_DEVICE_MIDIOUT__
    if(MIDIDeviceInit(&Devices[i].dev.MIDI)) {
      Devices[i].type = DEVICE_TYPE_MIDI;
      Devices[i].MIDIHandlers = &MIDIDevMIDIHandlers;
      SMInfo.MIDIDevice = i++;
      nPlaybackDevices++;
    }
  #endif
```

No `MIDIHANDLERS` are defined for the `__SLIB_DEVICE_MIDIIN__` device. Instead, we set up a function to be called whenever there's a MIDI message available on the selected input port. This happens when we activate the device in the `midiInOpen` call in `MIDIInDeviceActivate` (in `\Code\Sound\Win9x\MIDIDev.c`):

```
MIDIindevice *Mdev;
if(SMInfo.MIDIInDevice==SMINFO_NODEVICE) return(FALSE);
if(!Devices[SMInfo.MIDIInDevice].active) {
  Mdev = &Devices[SMInfo.MIDIInDevice].dev.MIDIIn;
  if(MMSYSERR_NOERROR == midiInOpen(&Mdev->hMIDIIn,
    Mdev->windowsDevID, (DWORD)MIDIInCallBack, 0,
    CALLBACK_FUNCTION | MIDI_IO_STATUS)) {
// Reset the MIDIIn device (start input)
if(MMSYSERR_NOERROR == midiInStart(Mdev->hMIDIIn)) {
  Devices[SMInfo.MIDIInDevice].active = TRUE;
  return(TRUE);
} else {
  //Error resetting MIDIIn device
    }
    midiInClose(Mdev->hMIDIIn);
  } else {
    //Error opening MIDIIn device
  }
  }
...
```

`MIDIInCallBack` places these messages into a static buffer, called `MIDIBuffer`, which we process during our timer interrupt. This happens in the `Service-MIDIBuffer` function, called from within `UpdateDevices` (see Chapter 9). We parse the incoming MIDI stream, and call the appropriate `MIDIHANDLERS` for each channel's `device`.

File it under MIDI

We complete our MIDI initialization by calling `InitMIDIFiles` from within `SMInit` (in `\Code\Sound\MIDIFile.c` and `\Code\Sound\SM.c`, respectively). This function is part of our `ModuleProcs` array (discussed in Chapter 4):

```
BOOLEAN InitMIDIFiles(void) {
  MIDIFiles = NULL;
  BeatAccumulator = 0;
  TicksPerFrame = 0;
  BeatModulo = 0;
  MIDIFileMarkerProc = NULL;
```

```
    FrameProportion = 1<<12;
    if(InitMasterPlayers()) {
      if(InitNoteTable()) {
      MasterVolumeProcListAdd(MusicMasterVolumePtr,
          (void*)MIDIFileSetMusicMasterVolume);
      return(TRUE);
    }
  ...
```

We set our global active `MIDIFiles` pointer to NULL, and initialize some timing values we'll use to synchronize external game behavior to our MIDI file playback. (We'll see how all this works in Chapter 11 when we talk about real-time coordination between the Soundtrack Manager and the game.) Here we can also set a public function pointer called `MIDIFileMarkerProc` to be called every time we encounter a Standard MIDI File (SMF) Marker Meta event message. The prototype for this callback, defined in `\Code\Sound\MIDIFile.h`, is:

```
typedef void
  (*MIDIFILEMARKERPROC)(MASTERPLAYER *masterPlayer,
    UINT32 markerNum, char *markerName);
```

This is another place where the game can be synchronized to the audio. For instance, a slide show or animation can be driven by strategically placed marker events in the MIDI file. The music lets you know where it is as it plays without any elaborate and often inaccurate timing schemes. The callback receives notification of the marker, and you choose what to do with that information. We'll give more details about this interactive hook below and in Chapters 9 and 10.

Continuing along in `InitMIDIFiles`, we call `InitMasterPlayers` to initialize our static linked list of `MASTERPLAYER`s, called `MasterPlayers`, to NULL. We will describe this structure in greater detail later in this chapter. For the moment, know that a `MASTERPLAYER` structure is what we use to keep track of and control our MIDI file playback.

Notes and volumes

This is followed by a call to `InitNoteTable`. The Soundtrack Manager keeps track of the MIDI channel, note, velocity, pan position, time off and `VOICEOWNER`s of each note that gets turned on. This allows us to independently manipulate the individual tracks of a MIDI file. As we'll see in Chapter 7, we can mute or unmute a track, or change its volume, pan position or program number, all on a note-by-note basis using the `NoteTable`. This is where we set that up.

Volume control is a complex issue for a sound engine. That's because there are so many different places where we control this important parameter. For instance, each digital audio sample has its own intrinsic volume, as does each individual MIDI note via its initial velocity. If we play back more than one SAMPLE or MIDI file, we want the facility to adjust their volumes independently of one another. SAMPLEs can be used as dialog. SAMPLEs and MIDI files can be used as either sound effects or as part of the music of a game. Each of these classes of sounds should have their own master volume adjustment. With the myriad possible combinations of absolute and relative levels, you can see why volume handling is a difficult task. Here in our InitMIDIFiles routine, we control the volume of all our MIDI files simultaneously by adding the MIDIFileSetMusicMasterVolume function to the Soundtrack Manager's global MusicMasterVolumePtr list of MASTERVOLUME pointers (defined in \Code\ Sound\Volume.h). We'll take up individual sound component volume issues when we present our cue-based text interface for audio artists later on in Chapter 10.

MIDI playback options

There are two ways in which the Soundtrack Manager plays MIDI. The first sends all MIDI commands to a hardware synthesizer located either on the soundcard or embedded processor of the machine. The advantages of doing this are that PC soundcards already have a General MIDI (GM) sound set installed, and you can get some noise out of the box without too much effort. A definite downside on the PC is that the GM sound set varies tremendously from soundcard to another, so you never know how your music is going to sound on the end user's system. Some platforms allow you to load custom samples into their hardware that are then triggered by these same MIDI commands. Using custom sample banks, you have a pretty good idea about how things are going to sound. The downside here is that not all machines have such cards.

The second way to play MIDI files within the Soundtrack Manager is to send all the commands to a software synthesizer using custom digital audio sound banks, or wavetables. If you happen to be working on the PC or XBox, this facility is built into those platforms in the form of DirectMusic. If you're not on one of those platforms, you have to roll your own (which we do in Chapter 7). In all of these cases, since the audio artist supplies the custom sample banks (the digital audio samples comprising the instruments themselves), the user reliably hears what the artist intended. It sounds the same no matter what physical machine you play it on, as it requires only the presence of a digital-to-analog converter. The disadvantages of this method is that it takes a lot of time to write a robust software synthesizer (or you could just buy a book about it), and more time to design and build compelling sound banks. Be very aware that this is not the same as simply ripping tracks from effects CDs. Creating good game audio content is an art and a craft practiced by many dedicated and talented professional individuals. I strongly encourage you to hire them.

The Soundtrack Manager provides a third option for playing MIDI. If the platform supports it, we can send those commands out to an external piece of MIDI equipment. I know of no game that allows you to do this (although I think it would be fun!), but it is extremely useful for debugging and testing one's MIDI handling code. Especially when comparing software synthesizer performance with a known piece of studio gear.

The Soundtrack Manager also allows you to define a device for interpreting incoming MIDI commands. Again, I know of no game that allows you to do this. However, this can be of enormous value to the audio artist while developing content for the game. He or she can load their sounds into an early version of the game or some other authoring application, and hear how their MIDI-controlled content is going to sound on the target platform itself. When time comes to ship the game, this support is removed from the Soundtrack Manager simply by recompiling the library without the __SLIB_DEVICE_MIDIIN__ definition.

Play it!

So, let's make some noise, already! How do I use the Soundtrack Manager to play back MIDI files? Read on:

```
MIDIFILE *midiFile = NULL;
MASTERPLAYER * myNewMasterPlayer = NULL;
if(midiFile=ReadNMDFile("MyFile.nmd"))
{
  if(myNewMasterPlayer=GetMasterPlayer(midiFile))
  {
    PlayMIDIFile(myNewMasterPlayer);
  }
}
```

This looks pretty straightforward. But what is a MIDIFILE, MASTERPLAYER and an NMD file? In the next few paragraphs, we reveal the MIDI wizard behind the curtain.

Parts is parts

The ReadNMDFile function (located in \Code\Sound\ReadNMD.c) takes a single argument, the name of the new MIDI file to open. Chapter 6 takes up this new MIDI file format in more detail. For now, understand that a '.nmd' file is a custom MIDI file format that supports looping, branching and arbitrary mid-file startup, something SMFs cannot do. All SMFs must be converted to NMD files before they can be used in our Soundtrack Manager. We provide a utility application to do this on the CD, called MIDI2NMD, and talk more about this SMF to NMD transformation in the Chapter 6. ReadNMDFile returns a pointer to a MIDIFILE structure, defined in \Code\Sound\MIDIFile.h as:

```
typedef struct _MIDIfile {
  UINT32      size;
  char        *name;
  UINT32      nTracks;
```

```
    MIDITRACK      *tracks;
    UINT32         pulsesPerQuarterNote;
    UINT32         endPulse;
    UINT32         uSecsPerQuarterNote;
    UINT8          timeSigNum;
    UINT8          timeSigDen;
    UINT16         markerEventNumber;
    struct _MIDIfile *next;
  } MIDIFILE;
```

As you can see, we keep track of a new MIDI file's `size`, `name` and number of tracks. The underlying time granule of an SMF, a pulse, is retained in our internal NMD file format. The number of `pulsesPerQuarterNote` is defined in its header, while the `uSecsPerQuarterNote` is calculated from the first tempo event of the file. We also store the first time signature event of the file in the `timeSigNum` and `timeSigDen` members. These values are used in conjunction with `pulsesPerQuarterNote` to calculate the number of `pulsesPerMeasure` in the `MASTERPLAYER` structure (described below).

The `MIDIFILE` structure stores the final pulse of the file for accurate looping in `endPulse`, and also holds a linked list of all the contained NMD tracks of this file. The structure of an NMD track is not shown here, but it can contain any number of markers to be used at the artist's and programmer's discretion. Markers are SMF Meta Event text messages that a composer can place in a MIDI file to signal a certain point in the music (e.g. first verse, refrain, coda, etc.). Support of markers provides another point where we can potentially control the overall flow of the game in response to the sound, or vice versa. As we'll see in Chapter 6, we can set up a function to be called whenever we hit a marker event. Using the `MIDIFILE`'s `markerEventNumber` member, we can specify a set of actions to perform when we hit the next marker. We'll talk about how we make this interactive audio mechanism work for audio artists in Chapter 10. For now, it is sufficient to know that all of the game audio programming services described in this book will be made available to the sound artist via a cue-based text interface.

The final member of our `MIDIFILE` structure is a pointer to another `MIDIFILE`. This allows us to create a global linked list of `MIDIFILE`s called, oddly enough, `MIDIFiles`. This list is used to keep track of all our allocated `MIDIFILE`s for easy releasing when we exit the game and close down our library.

Yes, master

Getting back to our MIDI file playback example, once we successfully read our NMD file, we call `GetMasterPlayer` to get a `MASTERPLAYER`. This is the structure we use to contain and actually play back a `MIDIFILE`, and is defined as:

```
  typedef struct _masterplayer {
    struct _MIDIfile *MIDIFile;
    struct _playlist *playList;
```

```
      struct _player *subPlayers;
      volatile BOOLEAN isPlaying;
      volatile BOOLEAN isPaused;
      volatile BOOLEAN isLooped;
      FIXED20_12      currentPulse;
      FIXED20_12      nextNoteTableTimeOff;
      FIXED20_12      pulsesPerFrame;
      UINT32          pulsesPerMeasure;
      FIXED20_12      currentMeasurePulse;
      UINT16          volume;
      INT16           pan;
      struct _fadeinfo *fade;
      struct _masterplayer *next;
   } MASTERPLAYER;
```

The first entry in this structure is a pointer to a _MIDIFile structure. You will notice that we do not use the type name MIDIFILE (in all caps) in this MASTERPLAYER structure definition. This is because we can never be sure what header files the compiler has processed by the time it gets around to this one (\Code\Sound\ MIDIFile.h). Therefore, those cool new type names may not be defined. This is not a problem at runtime, as all types have been defined by the time we try to execute anything. So to avoid any compiler "syntax error" messages, we often use the raw struct names, and not their new type names, in our structure definitions.

Following our MIDIFile member is a pointer to something we call a _playlist. As we'll see in Chapters 9 and 10, we can make a list of MIDIFILES to play one right after another – a PLAYLIST. We can perform pretty much any operation on PLAYLISTs that we can do on stand-alone MIDI files. In addition, we can append, remove, jump to or segue to any MIDI file contained in a particular PLAYLIST.

The next member of our MASTERPLAYER structure is a linked list of _player structures, called subPlayers. We use a _player to control the playback of an individual track of a MIDI file, and define it to be:

```
   typedef struct _player {
     MASTERPLAYER       *master;
     struct _MIDItrack  *track;
     volatile BOOLEAN   isMuted;
     BOOLEAN            isOver;
     UINT16             volume;
     INT16              pan;
     struct _fadeinfo   *fade;
     INT16              channel;
     FIXED20_12         nextNoteTableTimeOff;
     struct _player     *next;
     struct _proximitysoundlistentry *pSoundList;
   } PLAYER;
```

A PLAYER contains a pointer to its associated MASTERPLAYER, and a pointer to the _MIDItrack it controls. (We'll present more information about the MIDI track structure in Chapter 6 when we present the details of our new NMD MIDI file format.) Next we have a couple of flags to indicate if this track is muted or done playing, followed by the volume and pan members. As we saw for SAMPLEs, we allow each MIDI track, in the form of a PLAYER, to dynamically fade up or down over a specified period of time to a given volume and pan position through the use of a _fadeinfo structure.

The MIDI channel of this track is held in channel, and nextNoteTableTimeOff keeps track of the time of the next note off of this MIDI track. The next member of the PLAYER structure points to the next _player of this MASTERPLAYER's list of 'subPlayers.'

The final member of the PLAYER structure is a '_proximitysoundlistentry.' In Chapter 11, we'll see how this is used when we talk about real-time positional updates between the visual and auditory displays.

Continuing on with our MASTERPLAYER description, we define three boolean members to indicate whether the underlying MIDI file isPlaying, isPaused or isLooped. (I'm certainly feeling a bit looped as I go through all this stuff – but hang in there!). We keep track of our MIDI file's current play position in the currentPulse and currentMeasurePulse members. The time of the next note off event for this MIDI file is kept in nextNoteTableTimeOff (part of our NMD enhancements), along with two fields for adjusting the MIDI file's tempo in pulsesPerFrame and pulsesPerMeasure. The volume and pan fields are pretty self-explanatory, followed by our now-familiar fade element. As for SAMPLEs and PLAYERs, we can dynamically fade a MIDIFILE as a whole using this structure and our timer updates (in the UpdateMIDIFiles function, described below).

The last member of our MASTERPLAYER structure is a pointer to another MASTER-PLAYER. This allows us to create a global linked list of MASTERPLAYERs, called MasterPlayers. Similar to the Samples list of the previous chapter, we keep a linked list of all of our active MASTERPLAYER structures so we can easily tend to all of our current MIDI file audio resources during our periodic update interval.

MIDI activation

On platforms such as the PC, the user can have many different applications running simultaneously. She can also switch back and forth and among them at will. Any application, including a game, must release any shared audio system resources it may have claimed before it loses focus and is deactivated. It must also reclaim those resources when the game regains focus. In the Soundtrack Manager, just before the game loses focus, the game should call SMDeactivate (in \Code\Sound\SM.c). Here we perform any necessary deactivation operations on the underlying platform and deactivate all our defined devices. In the case of MIDI, for those channels not using the PCM device, SMDeactivate calls PlatformDeactivate. This sends an AllNotesOff MIDI controller message to each of our MIDI channels in the PlatformAllNotesOff function (in \Code\Sound\Win9x\SMPlatfm.c).

Any DirectSound secondary buffers that we may be using for MIDI are stopped and released in PCMDeviceAllNotesOff (in \Code\Sound\Win9x\DSMix\PCMDev.c).

SMDeactivate next calls DevicesDeactivate. If either the __SLIB_DEVICE _MIDIIN__ or __SLIB_DEVICE_MIDIOUT__ devices have been defined, we simply close those devices, and we're done. There is no additional deactivate function to call for MIDIFILE support.

When the game regains focus, the game should call SMActivate to activate the Soundtrack Manager. This calls ProcListExecuteB to call each of the functions in our ActivateProcList, followed by DevicesActivate (in \Code\Sound\ Win9x\Devices.c):

```
UINT32 DevicesActivate(void)
{
  UINT32 code = DEVICES_STARBOARDPOWERCOUPLING;
  if(DevicesInitialized()) {
    if(DevicesActive()) {
      return(DEVICES_OK);
    } else {
      SMOptions.platformInfo.devicesActive = FALSE;
#ifdef __SLIB_DEVICE_PCM__
      if(SMInfo.PCMDevice != SMINFO_NODEVICE) {
        code = PCMDeviceActivate();
        if(code != PCMDEVICE_OK) return(code);
      }
#endif
#ifdef __SLIB_DEVICE_MIDIOUT__
      if(SMInfo.MIDIDevice != SMINFO_NODEVICE) {
        if(!MIDIDeviceActivate())
          return(MIDIDEVICE_HARDWAREINUSEORABSENT);
      }
#endif
#ifdef __SLIB_DEVICE_MIDIIN__
      if(SMInfo.MIDIInDevice != SMINFO_NODEVICE) {
        MIDIInDeviceActivate();
      }
#endif
      SMOptions.platformInfo.devicesActive = TRUE;
      return(DEVICES_OK);
    }
  }
  return(FALSE);
}
```

We set our initial error code to DEVICES_STARBOARDPOWERCOUPLING (as in, "I can't 'old 'er, Cap'n!," for all you Trekkies out there). If our devices are already initialized and activated, we have nothing to do. For MIDI specifically, however, we call `MIDIDeviceActivate` if `__SLIB_DEVICE_MIDIOUT__` is defined, and `MIDIInDeviceActivate` if `__SLIB_DEVICE_MIDIIN__` is defined. These routines reopen the MIDI output and input devices chosen during `SMInit`. In `MIDIInDeviceActivate`, we also reset our MIDI input callback function to process incoming MIDI messages and restart the device (in `\Code\Sound\ Win9x\MIDIDev.c`).

Stop, start, pause and resume

We can stop or pause a playing `MIDIFILE` at any time by calling `StopMIDIFile` or `PauseMIDIFile`, respectively, specifying the appropriate `MASTERPLAYER` for the file. Both of these functions immediately stop the file's playback, but `Pause-MIDIFile` sets the `isPaused` `MASTERPLAYER` member to `TRUE` and turns off any notes that may be on. Any subsequent `PlayMIDIFile` or `ResumeMIDIFile` call will resume playback from the point the file was paused, including any events that were in mid-play when the `PauseMIDIFile` command came through.

   `StopMIDIFile`, on the other hand, sets both the `MASTERPLAYER` `isPaused` and `isPlaying` fields to `FALSE`. Any subsequent `PlayMIDIFile` call will consequently reset the play position to the beginning of the file before starting playback. A call to `ResumeMIDIFile` after calling `StopMIDIFile` will do nothing.

Quiet, already!

When we are done with our NMD file, we simply call `FreeNMDFile` (in `\Code\Sound\ ReadNMD.c`) and `FreeMasterPlayer` (in `\Code\Sound\MIDIFile.c`), as shown below:

```
//Recall that the MASTERPLAYER in our example
// was called, 'myNewMasterPlayer'...
MIDIFILE *midiFile = NULL;
if(myNewMasterPlayer) {
  if(midiFile=myNewMasterPlayer->MIDIFile) {
    FreeNMDFile(midiFile);
    FreeMasterPlayer(myNewMasterPlayer);
  }
}
```

These functions deallocate all the memory associated with their objects, remove them from their respective `MIDIFiles` and `MasterPlayers` linked lists, and free their specific data structures. As in the `FreeSample` routine in Chapter 4, all this is done inside a critical section to ensure all the housekeeping is finished before these lists are used or referenced by another thread.

We can find out whether a MIDIFILE is playing or paused by calling the MIDIFileGetPlayStatus function. This routine takes a pointer to the MASTER-PLAYER containing the MIDIFILE in question, and two boolean pointers to receive the playing and paused status we're interested in.

A MIDI file can contain any number of tracks. Once a MIDI file has been loaded into the Soundtrack Manager, MIDIFileGetNumTracks can be called to return the number of tracks a particular MIDI file has. This function takes only one argument, a pointer to the MASTERPLAYER in question. You can then find out the name of each track by passing MIDIFileTrackGetName a MASTERPLAYER pointer, the track number and a string to receive the track's name.

MIDI files and volume

As we've said before, in a game we have to be able to dynamically change the volumes of the sounds. This ability not only allows us to compensate for level differences among the sounds themselves, but also lets us set and change the audio mix during and in response to gameplay. To accommodate this dynamic behavior, the Soundtrack Manager API presents two functions to get and set the gain of an individual MIDIFILE. These are MIDIFileGetVolume and MIDIFileSetVolume. They both take a MASTERPLAYER pointer and a volume and pan parameter, and either get or set a MIDIFILE's volume and pan position. In MIDIFileSetVolume, if the file is currently playing it will adjust all the currently sounding notes within the mix. Otherwise, it just sets the MASTERPLAYER volume and pan members to their new values, affecting all upcoming notes when next the file is played.

The Soundtrack Manager also provides the same MIDIFILE volume facilities on an individual track basis. MIDIFileTrackGetVolume and MIDIFileTrack-SetVolume take a MASTERPLAYER pointer, the target track number within the MIDIFILE, and a volume and pan parameter. MIDIFileTrackSetVolume is used to adjust all the currently sounding notes within the mix. Otherwise, it just sets the individual tracks volume and pan to their new values, affecting all upcoming notes in that track when next the file is played.

The point is mute

Several API functions exist to allow you to mute or unmute a MIDIFILE or an individual track thereof. MIDIFileSetMute takes a pointer to a MASTERPLAYER, and an isMuted flag indicating which way you want to go. If isMuted is TRUE, MIDIFileSetMute traverses all the tracks of the specified file, and turns off any notes that may currently be playing. The isMuted member of each track's PLAYER object is also set to TRUE. Note that the MIDIFILE continues to play, even when muted. A subsequent call to MIDIFileSetMute with the isMuted parameter set to FALSE will pick up the file's playback from that point in time.

Muting and unmuting operations are not supported by Standard MIDI Files. All events in an SMF are stamped with delta times relative to the previous event. Therefore, unmuting an SMF, equivalent to restarting the file at an arbitrary point, involves racing through the file from the beginning to find the desired time and resuming playback from that point. Depending on the length of the file and at what point you want to restart, this could take quite awhile; that is, not immediately. In addition, any commands issued along the way that should still have an effect on the current music, such as pitch bends or program changes or notes that have not been turned off yet, are ignored. One could keep track of these events while furiously searching for the right moment in time, but there could easily be many such events along the way that have no bearing on the current music we're after. This is a waste of time and energy. There must be a better way!

Fortunately for us, there is. As we'll see in the next chapter, we've devised a new MIDI file format (NMD) that allows us to loop and branch and arbitrarily restart playback mid-file.

In the Soundtrack Manager API, there are two corollary functions to MIDI-FileSetMute that allow us to get and set the mute status of a MIDIFILE's individual tracks. MIDIFileTrackSetMute takes a pointer to a MASTERPLAYER, the track number upon which to operate, a flag indicating whether we want to mute the track or not, and what type of muting we want to perform. MIDIFILETRACK_MUTETYPE_INSTANT will perform the selected mute behavior immediately, while MIDIFILE-TRACK_MUTETYPE_NEXTEVENT will defer the operation until the next event in the MIDI file. MIDIFileTrackGetMute simply returns the muted status of the track about which we're inquiring.

I'm fading away

The MIDIFileSetFade function fades a MIDI file in or out to some target volume and pan position over a specified period of time. It takes as it's argument list a MASTERPLAYER pointer, the fade's target volume and pan, the number of milliseconds over which to perform the fade, and a flag to stop the sound playing at the end of the fade or not. This functionality is made possible by our underlying timer and update architecture via the UpdateMIDIFiles function, called every 8 ms. It is here that we turn off any notes that are less than our current pulse in UpdateNoteTable, and update our volumes and fades in UpdateMasterPlayers. If the fadeInfo field of the current MASTERPLAYER is defined, we call GetFadeVolume to calculate the fade's current volume and pan values and decide whether those parameters need to be updated in this frame or not. If so, we call MIDIFileSetVolume (in \Code\Sound\MIDIFile.c) to adjust the volume of all the currently playing notes, and check if we're done fading. At the end of the fade, we stop the MIDIFILE if the 'stopWhenDone' field of the FADEINFO structure is set. Otherwise, the sound continues to play at the target volume and pan position. Once the fade is complete, the 'fadeInfo' structure is cleared.

Miscellaneous MIDI

To loop a MIDI file, we call the `MIDIFileSetLoop` function. This function takes a `MASTERPLAYER` pointer and a flag to loop the file or not. Setting this flag to `TRUE` will loop the file from start to finish. Setting this flag to `FALSE` will disable MIDI file looping. The `MIDIFileGetLoop` function simply returns the value of the `isLooped` flag of the given `MASTERPLAYER`.

We can speed up or slow down the tempo of our MIDI files using the `MIDIFile-SetTempo` function. This takes a pointer to `MASTERPLAYER`, and a parameter to set the desired number of microseconds per quarter note, `uSecsPerQtrNote`. You can easily calculate the number of beats per minute (bpm) from `uSecsPerQtrNote`. For example, setting this parameter to 500 000 yields a bpm of 120:

```
bpm = (1000000 / uSecsPerQtrNote) * 60;
bpm = (1000000 / 500000) * 60;
bpm = 120;
```

Admittedly, this parameter is a bit funky, but used in conjunction with a MIDI file's pulses per quarter note value, we can set the new tempo (in `\Code\Sound\MIDIFile.c`). Each Soundtrack Manager timer interrupt is called a "frame." Therefore, our "`frameRate`" is the inverse of this timer period, 1.0/0.008, or 125 Hz. Every MIDI file contains a value called `pulsesPerQuarterNote` (ppq) that forms the basis of the delta time values between subsequent events in the file. For our internal tempo calculations, we have to transform this ppq value into the number of pulses per frame inside the Soundtrack Manager. This is done inside `MIDIFileSetTempo` using ppq and `uSecsPerQtrNote`. For example, setting ppq to 480, `uSecsPerQtrNote` to 480 000 and using a `frameRate` of 125, we calculate the equivalent pulses per frame to be 8:

```
pulsesPerFrame =
  ((ppq * 1000000) / uSecsPerQtrNote) / frameRate;
pulsesPerFrame = ((480 * 1000000) / 480000) / 125;
pulsesPerFrame = 8;
```

We store this value in the `MASTERPLAYER`'s `pulsesPerFrame`, and we're done. `MIDIFileGetTempo` simply returns the current `uSecsPerQtrNote` value of the `MASTERPLAYER`.

Calling markers

Earlier in this chapter, we talked about registering a procedure to be called every time we encounter a marker message in a MIDI file. This is done using the `MIDIFileSetMarkerProc` API call, passing a pointer to the function that is to be called. This function will receive a pointer to the MASTERPLAYER playing the MIDIFILE, the marker number and a pointer to a character string containing the marker name (given to it by the audio artist when preparing a script command for this feature – see Chapter 10). Given the MASTERPLAYER, you can make full use of the Soundtrack Manager's public MIDI API. This is another way the game can be synchronized to the audio, as anyone can set and use this communication channel, including the game itself. We put this mechanism to great use ourselves in Chapter 10 when we support English-language text scripts created by the audio artist.

Also inside a MIDI marker procedure, you can execute any of the scripts or cues alluded to in the preceding paragraph. We set up this behavior by default within our scripting facilities by registering the private function ScriptMIDIFileMarkerProc (in \Code\Sound\Script.c). That routine executes the cue, or event number, set by the MIDIFileSetMarkerEventNumber function within the public MIDIFILE API. This function takes a pointer to a MASTERPLAYER, and an event number identifying the specific cue to run when the next marker in this MIDIFILE is encountered.

There are two final routines in the Soundtrack Manager having to do with markers. The first is MIDIFileGetMarkerNum. This function returns the number of the marker, given its name. This marker number can then be used by the MIDIFile-SeekToMarker routine to seek to that point in the MIDIFILE.

Summary

This chapter detailed the implementation of another game audio asset – MIDI. The high-level musical services offered via MIDI were discussed, as well as some of the underlying support mechanisms. Platform-specific operations were again compartmentalized for easy replacing when moving to a different platform.

6 Case study 1: Extending MIDI file performance – start me up!

Games have traditionally used Standard MIDI Files (SMFs) to play back music and background tracks. But the SMF specification was never designed for interactive use, and therefore lacks support for such key behaviors as looping, branching or arbitrary mid-file startup. This case study describes the musical demands behind and the technical implementation of a custom MIDI file format and playback system for interactive games.

MIDI recap

In the last chapter, we talked about the use of the Musical Instrument Digital Interface, or MIDI, in games. By MIDI, we really meant playing SMFs. The MIDI specification, available from the MIDI Manufacturers Association (MMA), defines a set of commands for the real-time performance and control of music synthesizers. SMFs time-stamp and collect those commands into a file for easy interchange among different computers, akin to a digital score. They achieve tremendous compression ratios as they contain no audible information in and of themselves, but produce sound when their commands are streamed to a synthesizer. This along with their compact representation commonly yields a 200 : 1 compression ratio compared to a 44 kHz, 16-bit stereo recording of the synthesizer output. However, one challenge has always plagued MIDI: interactivity.

Standard MIDI files are consumed linearly. That is, you hit the start button, and out it comes. They only play from the beginning, and don't support any jumping around in the file. As mentioned in Chapter 4, games turn the notion of linear playback completely on its head. In an interactive game, each player creates his or her own audio experience as the game is played, and the game's soundtrack has to respond in an appropriate and timely way. Since we don't know a priori what actions the players will take, we can never know what sounds will follow one another. If SMF playback could be interactive, we could reap the space benefits they provide along while better serving the game and creating a more immersive audio experience.

MIDI becomes interactive

In the Soundtrack Manager, we have created a new interactive MIDI format called NMD. Based on the Standard MIDI File specification, this powerful new format supports file looping and interactive branching and gives us control over each MIDI track individually. It also supports arbitrary mid-file startup. Previous attempts to do this with SMFs were unsatisfactory, because they don't preserve any information about the past. Standard MIDI files store all of their event information in a monotonically increasing, time-relative manner. Therefore, they cannot recreate any particular musical performance state without starting over from the beginning. The NMD format allows us to do all of this, and requires no additional production tools. Standard MIDI files are converted to NMDs as a post-process before the content is inserted into the game. This allows audio artists to continue to use their sequencer of choice to author MIDI game content, but with these innovations, finally gain interactive control of their MIDI compositions.

In this chapter, we present and discuss the technical implementation of a custom MIDI file format and playback system. We'll discuss in detail inner workings of this new technology, and document the Soundtrack Managers extended MIDI API. Later on in Chapter 10, we'll present the final piece of this puzzle when we turn the power of the programming API over to the audio artists themselves via a simple English-language script interface. But for now, please refer to the MIDI2NMD project located on the CD (in the \Code\Apps\MIDI2NMD folder). MIDI2NMD is a Win32 console application to transform a SMF into our new MIDI file format, NMD.

What it is

At the top of the main.c file, we include MIDIConv.h. Within that header, we include the stypes.h and MIDIFile.h files to use the same type and MIDI file definitions as in the Soundtrack Manager. The application takes one argument, the name of the input MIDI file to be processed. You can either specify the name of the output file in a second, optional argument, or the program will create a new file to hold the NMD output, replacing the original files extension with .nmd.

The bulk of the work in this program takes in the MIDIConv.c source file. We make use of a number of Soundtrack Manager services in our conversion code, and so include several other header files at the top of this file: SMIO.h, SMMemMgr.h and MiscTool.h. The corresponding source files are also added to this project.

The application passes its first argument to the GetMIDIFile routine. This function opens the file, and validates that it is an SMF file. If it is, we allocate a new MIDIFILE structure, and begin to fill it in. (Please refer to Chapter 5 for a more in-depth look at a MIDIFILE):

```
if(hFile=SLib_FileOpen(fileName, SM_IO_OPEN_READ)) {
  if(SLib_FileRead(hFile, buffer, 4)==4) {
```

```
    if(strncmp(buffer, "MThd", 4)==0) {
      if(MIDIFile=(MIDIFILE*)SMGetMem(sizeof(MIDIFILE)))
        {
          APPMEMSET(MIDIFile, 0, sizeof(MIDIFILE));
          ...
```

We save the name of the file, the number of tracks it contains and the number of pulses per quarter note (`pulsesPerQuarterNote`, or `ppq`):

```
  MIDIFile->name = (char*)SMGetMem(strlen(fileName)+1);
    strcpy(MIDIFile->name, fileName);
    if(SLib_FileRead(hFile, buffer, 4)==4) {
      ckSize = GetALong(buffer);
      if(SLib_FileRead(hFile, buffer, ckSize)) {
          format = GetAWord(buffer);
          MIDIFile->nTracks = GetAWord(buffer+2);
          MIDIFile->pulsesPerQuarterNote =
                              GetAWord(buffer+4);
          status = TRUE;
      ...
```

It has a pulse

SMF events are stored in sequential, relative order. That is, each event specifies how long it is until the next event on the current track. These are unitless quantities called "pulses," and define the granularity of a MIDI file's time resolution. SMFs commonly use a value of 480 `ppq`, though it is not uncommon to set the `pulses-PerQuarterNote` to 960 or even higher if the file contains a number of very fast events. The real-time value of each pulse is set by the tempo events of the file, which specify the number of microseconds there are per quarter note (`uSecsPer-Quarter`). We can calculate the number of pulses per second at this tempo using the following expression:

```
  pulsesPerSecond = (pulsesPerQuarterNote * 1000000) /
                     uSecsPerQuarter;
```

This value, together with the Soundtrack Manager's frame rate, allows us to calculate how many pulses per frame there are for the `MIDIFILE`'s `MASTERPLAYER`:

```
  masterPlayer->pulsesPerFrame =
    pulsesPerSecond / frameRate;
```

For a frame rate of 125 Hz, corresponding to our 8 ms timer interrupt, a SMF that uses 480 `pulsesPerQuarterNote` and 480,000 `uSecsPerQuarter` yields

8 `pulsesPerFrame`. This means that every time our timer goes off, we increment our total file pulse counter by eight, and process any events that may have occurred in those last eight pulses.

First down

When we're through reading the SMF header, we sequentially read each track into memory. We allocate a `MIDITRACK` structure for each, and call `ParseMIDITrack` to read through and store all the events of a track:

```
for(trackNum=0; trackNum<MIDIFile->nTracks; trackNum++) {
  if(SLib_FileRead(hFile, buffer, 4)==4) {
    if(strncmp(buffer, "MTrk", 4)==0) {
      if(SLib_FileRead(hFile, buffer, 4)==4) {
        ckSize = GetALong(buffer);
          if(trkBuffer=(UINT8*)SMGetMem(ckSize)) {
          if(SLib_FileRead(hFile,trkBuffer,ckSize)==ckSize) {
            if(MIDITrack =
              (MIDITRACK*)SMGetMem(sizeof(MIDITRACK))) {
            APPMEMSET(MIDITrack, 0, sizeof(MIDITRACK));
              if(ParseMIDITrack(trkBuffer, ckSize, MIDITrack,
                          MIDIFile)) {
                . . .
```

The definition of a `MIDITRACK` is:

```
typedef struct _MIDItrack {
  char      *name;
  UINT32    nEvents;
  UINT8     *firstEvent;
  UINT8     *currentEvent;
  UINT32    currentEventNum;
  UINT32    currentPulse;
  UINT32    nextPulse;
  UINT32    nMarkers;
  struct    _MIDIfilemarker *marker;
  UINT32    seekPointInterval;
  UINT8     log2SeekPointInterval;
  UINT32    nSeekPoints;
  struct    _MIDIfileseekpoint *seekPoint;
  struct    _notetableentry *noteOnEvents;
  struct    _MIDItrack *next;
} MIDITRACK;
```

We start to get into the real nitty-gritty of the NMD file format with this structure. Each MIDITRACK stores the name of the track and the number of MIDI events it contains. The next few members let us know where we are in the file at any time. The firstEvent member contains the starting memory address of this track's event list, while currentEvent and currentEventNum contain the current event's address in our NMD RAM image and its number, respectively. The currentPulse and nextPulse fields keep track of the current and next event's pulse.

Standard MIDI files contain a number of special events, known as meta-events. These events, while generally not required, allow a composer to insert additional information into the file. We support four of these meta-events: end of track, set tempo, time signature and markers. A marker is a small bit of text describing the name of that point in the sequence: musically significant things such as a rehearsal letter or section name. We keep track of the number of such markers in each track, as well as a list of the marker events themselves (in nMarkers and marker, respectively). We'll detail the _MIDIfilemarker structure a little later on in this chapter.

The next four fields of our MIDITRACK structure have to do with seek points. These are the things that allow us to randomly jump into and around a MIDIFILE. As we'll see below, seekPointInterval and log2SeekPointInterval refer to the number of pulses between nSeekPoints spaced equidistantly throughout the track. We keep a list of these _MIDIfileseekpoint structures in the seekPoint member. Several public Soundtrack Manager API functions make direct use of these seek points, including MIDIFileSeekToMarker and MIDIFileSetMute.

The penultimate member of our MIDITRACK structure, noteOnEvents, is a list of all the currently sounding notes on this track. During MIDIFILE playback, we keep track of each note we turn on in a _notetableentry structure. This allows us to change the program, pitch, volume or pan position of any currently sounding notes individually on each track, as well as to mute or unmute tracks at will. The final member of our MIDITRACK structure is pointer to the next _MIDItrack.

Meanwhile, back at the ranch...

Back in ParseMIDITrack, we rip through each track, counting the number of events we have to save, and the total time (in pulses) of the track:

```
totalPulses = 0;
MIDITrack->nEvents = 0;
trkOffset = 0;
while(trkOffset < ckSize) {
  relativePulses = GetAVarLenQty(&trkBuffer[trkOffset],
                        &length);
  totalPulses += relativePulses;
  trkOffset += length;
  if(IsMTrkEventStatus(trkBuffer[trkOffset])) {
    runningStatus = trkBuffer[trkOffset++];
  }
```

```
if(IsMIDIEventStatus(runningStatus)) {
  MIDITrack->nEvents++;
  switch(runningStatus & 0xF0) {
  case NOTEON:
  case NOTEOFF:
  case POLYPRESSURE:
  case CONTROL:
  case PITCHBEND:
    trkOffset+=2;
    break;
  case PROGRAM:
  case AFTERTOUCH:
    trkOffset++;
    break;
  }
} else if(IsSysexEventStatus(runningStatus)) {
// We ignore sysex events
msgLength = GetAVarLenQty(&trkBuffer[trkOffset],
                         &length);
trkOffset += length + msgLength;
} else if(IsMetaEventStatus(runningStatus)) {
metaEventType = trkBuffer[trkOffset++];
msgLength = GetAVarLenQty(&trkBuffer[trkOffset],
                         &length);
trkOffset += length;
switch (metaEventType) {
  case MARKER:
    MIDITrack->nEvents++;
    MIDITrack->nMarkers++;
    break;
  case END_OF_TRACK:
    MIDITrack->nEvents++;
    break;
  case SET_TEMPO:
    // Don't bother making an event for the first
    // tempo event;it'll go in the MIDIFile's data
    if(numTempoEvents) {
      MIDITrack->nEvents++;
    }
    numTempoEvents++;
    break;
  case TIME_SIGNATURE:
    MIDITrack->nEvents++;
    break;
  default:
```

```
        break;
    }
    trkOffset += msgLength;
  }
}
```

Second down

We allocate memory for all these MIDIFILEEVENTs and space for a list of MIDIFILEMARKERs as well. We then scan through the tracks events a second time, filling in these structures:

```
if(MIDITrack->firstEvent = SMGetMem(MIDITrack->nEvents *
            sizeof(MIDIFILEEVENT))) {
  if(MIDITrack->nMarkers) {
    MIDITrack->marker = SMGetMem(MIDITrack->nMarkers *
            sizeof(MIDIFILEMARKER));
  } else {
    MIDITrack->marker = NULL;
  }
  trkOffset = 0;
  eventNum = 0;
  markerNum = 0;
  currentPulse = 0;
  while(trkOffset < ckSize) {
    relativePulses=GetAVarLenQty(&trkBuffer[trkOffset],
                    &length);
    currentPulse += relativePulses;
    trkOffset += length;
    ...
```

New events

The MIDIFILEEVENT structure is defined in MIDIFile.h to be:

```
typedef struct _MIDIfileevent {
  UINT8    status;
  UINT32   start;
  UINT32   duration;
  union {
    SMMIDIEVENT    MIDIEvent;
    METAEVENT      metaEvent;
```

```
    } uEvt;
    struct _player *player;
    BOOLEAN (*HandleEvent)(struct _MIDIfileevent
                           *MIDIFileEvent);
} MIDIFILEEVENT;
```

We use this structure to store both regular MIDI events and meta-events. The
status member tells us the MIDI event contained in this structure. start defines
the starting pulse of this event, while duration tells us the length of this event, also
in pulses. In another pass through our list of events later on, we'll calculate the actual
duration of each event where appropriate. This time through we set all note durations
to MIDIFILEEVENT_DURATION_INFINITE.

```
if(IsMTrkEventStatus(trkBuffer[trkOffset])) {
  runningStatus = trkBuffer[trkOffset++];
}
if(IsMIDIEventStatus(runningStatus)) {
  MIDIFileEvent =
  (MIDIFILEEVENT*)(MIDITrack->firstEvent +
             (eventNum * sizeof(MIDIFILEEVENT)));
  eventNum++;
  MIDIFileEvent->start = currentPulse;
  MIDIFileEvent->duration = MIDIFILEEVENT_DURATION_INFINITE;
  MIDIFileEvent->status = runningStatus & 0xF0;
  MIDIFileEvent->HandleEvent = NULL;
  ...
```

The final two members of the MIDIFILEEVENT structure are defined at runtime.
player points to a _player structure that is used to play back a particular
MIDITRACK, and HandleEvent is a function pointer to the specific code to handle
the given MIDIFILEEVENT.

A union of two event structures, SMMIDIEVENT and METAEVENT, follows next
in uEvt. MIDIEvent contains a MIDIMESSAGE, which is itself a union of all the
standard MIDI channel voice messages (note on, note off, control change, program
change, etc.). metaEvent is a METAMESSAGE that is a union of all the SMF meta-
event messages:

```
typedef struct _MIDIevent {
  MIDIMESSAGE MIDIMessage;
} SMMIDIEVENT;

typedef struct _MIDImessage {
  UINT8 channel;
  union {
    NOTEOFFMSG noteoff;
```

```
    NOTEONMSG noteon;
    POLYPRESSUREMSG polypressure;
    CONTROLMSG control;
    PROGRAMMSG program;
    AFTERTOUCHMSG aftertouch;
    PITCHBENDMSG pitchbend;
  } uMsg;
} MIDIMESSAGE;

typedef struct _metaevent {
  UINT8 type;
  METAMESSAGE metaMessage;
} METAEVENT;

typedef union {
  SEQUENCENUMBERMSG   sequenceNumber;
  TEXTEVENTMSG        textEvent;
  TEXTEVENTMSG        copyRight;
  TEXTEVENTMSG        trackName;
  TEXTEVENTMSG        instrumentName;
  TEXTEVENTMSG        lyric;
  MARKERMSG           marker;
  TEXTEVENTMSG        cuePoint;
  CHANNELPREFIXMSG    channelPrefix;
  SETTEMPOMSG         setTempo;
  SMPTEOFFSETMSG      SMPTEOffset;
  TIMESIGNATUREMSG    timeSignature;
  KEYSIGNATUREMSG     keySignature;
  CUSTOMMSG           custom;
} METAMESSAGE;
```

All of the messages associated with these SMMIDIEVENT and METAEVENT structures can be found in \Code\Sound\Mdef.h.

Back in the code (line 280 of MIDIConv.c):

```
MIDIEvent = &MIDIFileEvent->uEvt.MIDIEvent;
channel = MIDIEvent->MIDIMessage.channel =
            (runningStatus & 0x0F);
switch(MIDIFileEvent->status) {
case NOTEON:
  note = trkBuffer[trkOffset++];
  vel = trkBuffer[trkOffset++];
  if(vel > 0) {
    MIDIEvent->MIDIMessage.uMsg.noteon.note = note;
    MIDIEvent->MIDIMessage.uMsg.noteon.vel = vel;
```

```
    } else {
      MIDIFileEvent->status = NOTEOFF;
      MIDIEvent->MIDIMessage.uMsg.noteoff.note = note;
      MIDIEvent->MIDIMessage.uMsg.noteoff.vel = 0;
    }
    break;
  case NOTEOFF:
  ...
```

If we don't have a MIDI channel voice message, we ignore any MIDI system exclusive events, and check if we have a meta event:

```
  } else if(IsSysexEventStatus(runningStatus)) {
    // We ignore sysex events
    msgLength = GetAVarLenQty(&trkBuffer[trkOffset],
                              &length);
    trkOffset += length + msgLength;
  } else if(IsMetaEventStatus(runningStatus)) {
    metaEventType = trkBuffer[trkOffset++];
    msgLength = GetAVarLenQty(&trkBuffer[trkOffset],
                              &length);
    trkOffset += length;
    switch (metaEventType) {
      case SEQUENCE_NAME:
        if(MIDITrack->name=(char*)SMGetMem(msgLength+1)) {
          strncpy(MIDITrack->name, &trkBuffer[trkOffset],
                  msgLength);
          MIDITrack->name[msgLength] = '\0';
      }
      break;
  case MARKER:
  ...
```

The definition of a MIDIFILEMARKER is:

```
typedef struct _MIDIfilemarker {
  UINT8   *currentEvent;
  char    *name;
  UINT32  pulse;
} MIDIFILEMARKER;
```

currentEvent points to this marker's MIDIFILEEVENT, while name contains the text of the marker itself. We set pulse to the MIDI pulse of the event in this MIDITRACK:

```
MIDIFileEvent = (MIDIFILEEVENT*)(MIDITrack->firstEvent +
                (eventNum * sizeof(MIDIFILEEVENT)));
eventNum++;
```

```
MIDIFileEvent->start = currentPulse;
MIDIFileEvent->duration = 0;
MIDIFileEvent->status = runningStatus;
metaEvent = &MIDIFileEvent->uEvt.metaEvent;
metaEvent->type = metaEventType;
metaEvent->metaMessage.marker.markerNum = markerNum;
// now fill in the MIDIFILEMARKER struct
if(marker = &(MIDITrack->marker[markerNum++])) {
  if(marker->name=(char*)SMGetMem(msgLength+1)) {
    strncpy(marker->name, &trkBuffer[trkOffset],
            msgLength);
    marker->name[msgLength] = '\0';
  }
  marker->pulse = currentPulse;
  marker->currentEvent = NULL;
}
break;
...
```

MIDI snapshots

If you could take a snapshot of a SMF track's playback at any moment, you would see some number of notes playing on any number of instruments influenced by any number of pitch bend, control or tempo messages. However, you would have no way of knowing how you got to this particular play state, as those events that occurred before the snapshot was taken are gone forever. With a traditional MIDI file, there is no way to recreate this play state but to start over from the beginning of the file and play until that same moment in time. This wouldn't be too hard if the point you were interested in were near the start of the file. But this becomes increasingly burdensome and time-consuming as that point moves further into the file and as the file length increases. It would be very nice, not to mention extremely useful, if we could start a MIDI file playing at an arbitrary point, with a minimum of hassle and chasing. This is exactly the impetus behind what we call a MIDIFILESEEKPOINT. In this structure, we record all the events that create the current state of the MIDI sequence at a given moment in time. By placing these seek points evenly throughout our new NMD file, random access startup becomes an efficient reality.

When we receive a request to begin playing our MIDI file at a given pulse, we go to the closest MIDIFILESEEKPOINT before that pulse and process all the events in that seek point, called CATCHUPEVENTs. We still have to scan through the file from that point and process all the events from the seek point pulse to the desired pulse. But if we put in enough seek points, we can always recreate any point in the music quickly and easily:

```
typedef struct _MIDIfileseekpoint {
  UINT32      pulse;
```

Audio Programming for Interactive Games

```
UINT8        *currentEvent;
UINT32       currentEventNum;
UINT32       refPulse;
UINT32       nCatchUpEvents;
CATCHUPEVENT *catchUpEvent;
} MIDIFILESEEKPOINT;
typedef struct _catchupevent
{
UINT8    *event;
UINT32   refPulse;
} CATCHUPEVENT;
```

We set `pulse` to the MIDI `pulse` of each `MIDIFILESEEKPOINT`, and point to the current `MIDIFILEEVENT` that is this seek point.

Back to our MIDI file processing, we make an initial guess at the seek point interval spacing in line 415 of `MIDIConv.c` (in `ParseMIDITrack`). To make our track searches faster, we quantize this interval to the closest power of two pulses less than or equal to our initial pulse spacing estimate. We set the number of seek points by dividing the total number of pulses by this new seek point interval, and allocate and initialize an array of `MIDIFILESEEKPOINT`s. The `pulse` member of each seek point is set to the current seek point number times the newly quantized seek point interval:

```
MIDITrack->seekPointInterval =
  MAX(1, MAX_MIDIFILEEVENTS_PER_INTERVAL * totalPulses /
        MAX(1, MIDITrack->nEvents));
  QuantizeSeekPointInterval(
    &MIDITrack->seekPointInterval,
    &MIDITrack->log2SeekPointInterval);
MIDITrack->nSeekPoints =
  (totalPulses / MIDITrack->seekPointInterval) + 1;
// Here we an array of markers, and set
// MIDITrack->seekPoint to the address of the first
// element in the array
if(MIDITrack->seekPoint =
  (MIDIFILESEEKPOINT*)SMGetMem(MIDITrack->nSeekPoints*
                       sizeof(MIDIFILESEEKPOINT))) {
  for(currentSeekPoint=0;
    currentSeekPoint < MIDITrack->nSeekPoints;
    currentSeekPoint++) {
  seekPoint =
    &(MIDITrack->seekPoint[currentSeekPoint]);
  seekPoint->pulse = currentSeekPoint *
                        MIDITrack->seekPointInterval;
  seekPoint->currentEvent = NULL;
  seekPoint->nCatchUpEvents = 0;
```

```
        seekPoint->catchUpEvent = NULL;
    }
} else {
  return(FALSE);
}
...
```

Third down

On our third traversal of the events of this MIDI file, we read the MIDIFILEEVENTs we just created above instead of the input MIDI file. We keep running tables of all the note on, controller, program and pitch bend messages and use these to set the durations of all the events in our NMD file:

```
for(channel=0; channel<NUM_MIDI_CHANNELS; channel++) {
  for(note=0; note<NUM_MIDI_NOTES; note++) {
    noteOnTable[channel][note].event = NULL;
  }
  programTable[channel].event = NULL;
  for(controller=0; controller<NUM_MIDI_CONTROLLERS;
        controller++) {
    controlTable[channel][controller].event = NULL;
  }
  pitchBendTable[channel].event = NULL;
}
lastTempoEvent.event = NULL;
lastTimeSigEvent.event = NULL;
currentSeekPoint = 0;
currentPulse = 0;
for(eventNum=0; eventNum < MIDITrack->nEvents;
        eventNum++) {
MIDIFileEvent = (MIDIFILEEVENT*) (MIDITrack->firstEvent
                + (eventNum * sizeof(MIDIFILEEVENT)));
currentPulse = MIDIFileEvent->start;
// now generate the event durations
if(IsMIDIEventStatus(MIDIFileEvent->status)) {
  MIDIEvent = &(MIDIFileEvent->uEvt.MIDIEvent);
  channel = MIDIEvent->MIDIMessage.channel;
  switch(MIDIFileEvent->status) {
    case NOTEON:
      note = MIDIEvent->MIDIMessage.uMsg.noteon.note;
      if(MIDIEvent->MIDIMessage.uMsg.noteon.vel > 0) {
        if(noteOnTable[channel][note].event) {
          noteOnTable[channel][note].event->duration=
```

```
(currentPulse-noteOnTable[channel] [note].event->start);
        }
        noteOnTable[channel] [note].event =
          MIDIFileEvent;
      } else {
        if(noteOnTable[channel] [note].event) {
          noteOnTable[channel] [note].event->duration=
(currentPulse - noteOnTable[channel] [note].event->start);
        noteOnTable[channel] [note].event = NULL;
    }
  }
  break;
case NOTEOFF:
...
```

Fourth down

Finally in `ParseMIDITrack`, we count the number of catch-up events we'll need for
each seek point in one final run through of our `MIDIFILEEVENT`s:

```
for(eventNum=0; eventNum<MIDITrack->nEvents; eventNum++) {
  MIDIFileEvent = (MIDIFILEEVENT*)(MIDITrack->firstEvent
                  + (eventNum * sizeof(MIDIFILEEVENT)));
  currentPulse = MIDIFileEvent->start;
  // Count the number of catchUpEvents needed for each
  // seekpoint. The case statement is necessary because
  // we don't want to calculate for a seekpoint that
  // would be positioned at an event that we're throwing
  // away, like a noteOff or polyPressure.
  if(IsMIDIEventStatus(MIDIFileEvent->status)) {
    switch(MIDIFileEvent->status) {
    case NOTEON:
    case CONTROL:
    case PROGRAM:
    case PITCHBEND:
    eventIsAKeeper = TRUE;
    break;
  default:
    eventIsAKeeper = FALSE;
    break;
  }
} else if(IsMetaEventStatus(MIDIFileEvent->status)) {
  metaEvent = &(MIDIFileEvent->uEvt.metaEvent);
  switch (metaEvent->type) {
```

```
      case MARKER:
      case END_OF_TRACK:
      case SET_TEMPO:
      case TIME_SIGNATURE:
        eventIsAKeeper = TRUE;
        break;
      default:
        eventIsAKeeper = FALSE;
        break;
      }
    }
    if(eventIsAKeeper) {
      while(currentSeekPoint <= (currentPulse >>
              MIDITrack->log2SeekPointInterval)) {
        seekPoint =
          &(MIDITrack->seekPoint[currentSeekPoint]);
        seekPoint->currentEventNum = eventNum;
        prevEventListEntry = NULL;

        //Traverse all events in eventList looking for
        // those whose end times are beyond the current
        // pulse of this seek point
        for(eventListEntry=eventList; eventListEntry;
          eventListEntry=nextEventListEntry) {
        nextEventListEntry = eventListEntry->next;
        if(eventListEntry->end > currentPulse) {
          seekPoint->nCatchUpEvents++;
          prevEventListEntry = eventListEntry;
        } else {
          //Remove the current eventListEntry from the
          // eventList (end <= currentPulse)
          if(prevEventListEntry) {
            prevEventListEntry->next =
              eventListEntry->next;
          } else {
            eventList = eventListEntry->next;
          }
          SMFreeMem(eventListEntry);
        }
      }
        currentSeekPoint++;
    }

        //Build linked list of events, keeping track of
        // their ending MIDI pulse
```

```
    if(eventListEntry =
 (MFEVENTLISTENTRY*)SMGetMem(sizeof(MFEVENTLISTENTRY)) {
     eventListEntry->event = MIDIFileEvent;
      if(MIDIFileEvent->duration == 0xFFFFFFFF) {
       eventListEntry->end = MIDIFileEvent->duration;
      } else {
       eventListEntry->end = MIN((currentPulse +
                MIDIFileEvent->duration), 0xFFFFFFFF);
      }
      eventListEntry->next = eventList;
      eventList = eventListEntry;
    }
   }
  }
  return(TRUE);
```

GetMIDIFile **next calls** `MIDIFileToNMDFile` **to complete the transformation (in line 139 of** `MIDIConv.c`):

```
if(NMDFile = MIDIFileToNMDFile(MIDIFile)) {
  FreeTransitionalMIDIFile(MIDIFile);
  return(NMDFile);
}
```

At this point, we have constructed a `MIDIFILE` in memory including a number of allocations for `MIDITRACK`s, `MIDIFILEEVENT`s, `MIDIFILEMARKER`s and `MIDIFILE SEEKPOINT`s. This function consolidates those discontinuous portions into a single RAM image of a new `NMD` file.

Making the NMD file

The first thing we do is to calculate how large the final output file will be. Once we know this size, we allocate one large contiguous block of memory to contain it. For each track in our `MIDIFILE`, we copy all the events of those tracks into the `MIDIFileImage` we're building:

```
//Set the size of the output NMD file, and fill in
// the RAM image of the file
MIDIFile->size = size;
offset = 0;
if(MIDIFileImage=SMGetMem(MIDIFile->size)) {
memcpy(MIDIFileImage, MIDIFile, sizeof(MIDIFILE));
offset += sizeof(MIDIFILE);
if(MIDIFile->name) {
```

```
  strcpy(MIDIFileImage+offset, MIDIFile->name);
  offset +=
  (((strlen(MIDIFile->name)+1)+3) & 0xFFFFFFFC);
}
trackNum = 0;
for(MIDITrack=MIDIFile->tracks; MIDITrack;
      MIDITrack=MIDITrack->next, trackNum++) {
 memcpy(MIDIFileImage+offset, MIDITrack,
        sizeof(MIDITRACK));
 offset += sizeof(MIDITRACK);
 if(MIDITrack->name) {
   strcpy(MIDIFileImage+offset, MIDITrack->name);
   offset +=
     (((strlen(MIDITrack->name)+1)+3) & 0xFFFFFFFC);
 }
 lastMIDIFileEvent = NULL;
 for(eventNum = 0; eventNum < nBigEvents[trackNum];
        eventNum++) {
   MIDIFileEvent =
   (MIDIFILEEVENT*)(MIDITrack->firstEvent +
     (eventNum * sizeof(MIDIFILEEVENT)));
 if(lastMIDIFileEvent) {
   deltaTime = MIDIFileEvent->start -
        lastMIDIFileEvent->start;
 } else {
   deltaTime = MIDIFileEvent->start;
 }
 duration = MIDIFileEvent->duration;
 if(IsMIDIEventStatus(MIDIFileEvent->status)) {
    channel =
MIDIFileEvent->uEvt.MIDIEvent.MIDIMessage.channel;
   switch(MIDIFileEvent->status) {
   case NOTEON:
     lastMIDIFileEvent = MIDIFileEvent;
     type = NMDEVENT_TYPE_NOTE |
   GetVariableTimeSizeCodes(deltaTime, duration);
     data1 = MIDIFileEvent->
     uEvt.MIDIEvent.MIDIMessage.uMsg.noteon.note;
     data2 = MIDIFileEvent->
     uEvt.MIDIEvent.MIDIMessage.uMsg.noteon.vel;
     memcpy((MIDIFileImage+offset++), &type, 1);
     memcpy((MIDIFileImage+offset++),
            &channel, 1);
     memcpy((MIDIFileImage+offset++), &data1, 1);
     memcpy((MIDIFileImage+offset++), &data2, 1);
```

```
        WriteVariableTimes(MIDIFileImage, &offset,
                            deltaTime, duration);
        break;
    case CONTROL:
    ...
```

After all the events, we copy the markers, and determine the addresses of the individual seek points of that track:

```
if(MIDITrack->nMarkers) {
  marker=&(MIDITrack->marker[0]);
  memcpy(MIDIFileImage+offset, marker,
    (MIDITrack->nMarkers * sizeof(MIDIFILEMARKER)));
  offset += (MIDITrack->nMarkers *
            sizeof(MIDIFILEMARKER));
  for(markerNum=0; markerNum<MIDITrack->nMarkers;
        markerNum++) {
    marker=&(MIDITrack->marker[markerNum]);
    if(marker->name) {
      strcpy(MIDIFileImage+offset, marker->name);
      offset +=
        (((strlen(marker->name)+1)+3) & 0xFFFFFFFC);
    }
  }
}
seekPoint=&(MIDITrack->seekPoint[0]);
memcpy(MIDIFileImage+offset, seekPoint,
  (MIDITrack->nSeekPoints * sizeof(MIDIFILESEEKPOINT)));
offset += (MIDITrack->nSeekPoints *
          sizeof(MIDIFILESEEKPOINT));
for(seekPointNum=0;
  seekPointNum < MIDITrack->nSeekPoints;
    seekPointNum++) {
  seekPoint = &(MIDITrack->seekPoint[seekPointNum]);
  offset += (seekPoint->nCatchUpEvents *
              sizeof(CATCHUPEVENT));
}
```

The last routine of our `MIDI2NMD` conversion is `GetNMDFileFromImageMC`:

```
if(NMDFile=GetNMDFileFromImageMC(MIDIFileImage)) {
  return(NMDFile);
}
```

They call me the seeker

After resetting all our name and event fields to point within our NMD RAM image, we make one final pass through our MIDI events, now in our MIDIFileImage, to build the CATCHUPEVENTs of each seek point. As we move through the image, we construct a temporary list of all the events, saving a pointer to the event itself as well as the starting time and duration of each. This may seem redundant, as we already have that information in our MIDIFILE image. The difference is that as we get deeper into the file, we occasionally hit a seek point. When this happens, we run through our accumulated event list looking for any events whose end time exceeds the current seek points pulse. Each one that we find is saved in our seek point's catch-up event list. If the event is already over, we remove it from our event list, and continue on:

```
markerNum       = 0;
seekPointNum    = 0;
currentPulse    = 0;
lastPulse       = 0;
if(MIDITrack->nEvents > 0) {
  offset = (MIDITrack->firstEvent - MIDIFileImage);
  eventList = NULL;
  for(eventNum=0; eventNum<MIDITrack->nEvents;
      eventNum++) {
    MIDIFileEvent = (MIDIFileImage + offset);
    type = *MIDIFileEvent;
    offset += GetDataSizeFromCode(type);
    deltaTimeSize = GetDeltaTimeSizeFromCode(type);
    deltaTime =
        ReadVariableSize((MIDIFileImage+offset),
                                        deltaTimeSize);
    lastPulse = currentPulse;
    currentPulse += deltaTime;
    offset += deltaTimeSize;
    durationSize = GetDurationSizeFromCode(type);
    duration =
        ReadVariableSize((MIDIFileImage+offset),
                          durationSize);
    offset += durationSize;
    while(seekPointNum <=
  (currentPulse >> MIDITrack->log2SeekPointInterval)) {
      seekPoint =
            &(MIDITrack->seekPoint[seekPointNum]);
      seekPoint->currentEvent = MIDIFileEvent;
      seekPoint->currentEventNum = eventNum;
      seekPoint->refPulse = lastPulse;
```

```
//This is where each seek point is built,
// starting with the state according to the
// event list.
catchUpEventNum = 0;
prevEventListEntry = NULL;
for(eventListEntry=eventList; eventListEntry;
        eventListEntry=nextEventListEntry) {
  nextEventListEntry = eventListEntry->next;
  if(eventListEntry->end > currentPulse) {
    catchUpEvent =
&(seekPoint->catchUpEvent[catchUpEventNum++]);
    catchUpEvent->event =
                eventListEntry->event;
    catchUpEvent->refPulse =
                eventListEntry->refPulse;
    prevEventListEntry = eventListEntry;
  } else {
    //This event is over; take out of list
    if(prevEventListEntry) {
      prevEventListEntry->next =
          eventListEntry->next;
    } else {
      eventList = eventListEntry->next;
    }
    SMFreeMem(eventListEntry);
  }
}
if(catchUpEventNum !=
        seekPoint->nCatchUpEvents) {
  //Error - missing CatchUpEvents for this
  // seek point
}
  seekPointNum++;
}
  //Now set up the tables that store all this
  // seekPoint data as well as generate the end
  // times for events
switch(type & 0xF0) {
case NMDEVENT_TYPE_NOTE:
case NMDEVENT_TYPE_CONTROL:
case NMDEVENT_TYPE_PROGRAM:
case NMDEVENT_TYPE_PITCHBEND:
case NMDEVENT_TYPE_SETTEMPO:
case NMDEVENT_TYPE_TIMESIGNATURE:
  if(eventListEntry =
```

```
(EVENTLISTENTRY*)SMGetMem(sizeof(EVENTLISTENTRY)){
    eventListEntry->event = MIDIFileEvent;
    eventListEntry->refPulse = lastPulse;
    if(duration == 0xFFFFFFFF) {
      eventListEntry->end = duration;
    } else {
      eventListEntry->end =
      MIN((currentPulse+duration), 0xFFFFFFFF);
    }
    eventListEntry->next = eventList;
    eventList = eventListEntry;
  }
  break;
case NMDEVENT_TYPE_MARKER:
  marker = &(MIDITrack->marker[markerNum++]);
  marker->currentEvent = MIDIFileEvent;
  break;
default:
  break;
  }
}
```

That's it! We've done it! After saving the above NMD file image, we have a MIDI file that supports random time and marker-based file seeking, as well as musically appropriate mid-file startup. It was a bit painful, but we're now enormously rewarded by the increased interactivity of this vital resource. Let's take a closer look at how we use this new NMD format, and what it buys us.

Playing NMD files

You will recall that we periodically update our Soundtrack Manager by calling the SoundManager routine in \Code\Sound\SM.c. Within this function, we execute our UpdateProcList to update all of our defined audio modules. If we defined __SLIB_MODULE_MIDIFILE__, this includes the UpdateMIDIFiles function in \Code\Sound\MIDIFile.c. This calls UpdateMasterPlayers to service all of our currently playing MIDIFILEs, each encapsulated in a MASTERPLAYER stored in a list. This calls UpdateMasterPlayer where the real work happens. Each track of a MIDIFILE is assigned to a PLAYER structure and stored in a list of PLAYERs inside the MASTERPLAYER. Beginning in line 1237 of MIDIFile.c, we check if the time-stamp of the next event on the track is less than or equal to the current pulse of the MASTERPLAYER. If it is, then we call BuildMIDIFileEvent to process that event:

```
if((player->track->nextPulse << 12) <=
    player->master->currentPulse) {
  while(BuildMIDIFileEvent(&MIDIFileEvent,
```

```
        player->track->currentEvent,
        player->track->currentPulse) &&
        ((MIDIFileEvent.start << 12) <=
          player->master->currentPulse)) {
        ...
```

This function converts the stored event in our NMD file into a full MIDIFILEEVENT structure. It copies the data from the NMD memory image into the appropriate MIDIFILEEVENT fields, and sets the HandleEvent function pointer to the specific MIDI event handler for this event:

```
//"data" = player->track->currentEvent
if(data) {
  type = *data;

  dataSize = GetDataSizeFromCode(type);
  channel = *(data+1);
  data1    = *(data+2);
  if(dataSize > 3) {
    data2 = *(data+3);
  }
  switch(type & 0xF0) {
  case NMDEVENT_TYPE_NOTE:
    MIDIFileEvent->uEvt.MIDIEvent.MIDIMessage.channel=
      channel;
    MIDIFileEvent->
      uEvt.MIDIEvent.MIDIMessage.uMsg.noteon.note =
        data1;
    MIDIFileEvent->
      uEvt.MIDIEvent.MIDIMessage.uMsg.noteon.vel =
        data2;
    MIDIFileEvent->HandleEvent =
      HandleMIDIFileEventNoteOn;
    break;
  case NMDEVENT_TYPE_CONTROL:
  ...
```

Handling events

Upon returning from BuildMIDIFileEvent, we set the player member of the MIDIFILEEVENT to the PLAYER of this track, and call the recently set HandleEvent function pointer:

```
currentPulseBeforeEvent = player->master->currentPulse;
MIDIFileEvent.player = player;
```

```
player->track->currentPulse = MIDIFileEvent.start;
if(MIDIFileEvent.HandleEvent) {
  MIDIFileEvent.HandleEvent(&MIDIFileEvent);
}
```

In the case of channel voice messages, the underlying MIDIMESSAGE is extracted from the event in the HandleMIDIFileEventX functions (in \Code\Sound\ MIDIFile.c), and the MIDI handler for the specific device is called.

What? Another MIDI handler? What gives? This is where we make the transition from Soundtrack Manager operation to platform-specific services.

Way back when the Soundtrack Manager was started, when the application called the SMInit function, we called InitDevices which in turn called InitMIDI-Handlers (in \Code\Sound\Win9x\Devices.c). This routine called MIDI-DeviceMIDIHandlersInit and PCMDeviceMIDIHandlersInit, depending on the audio module definitions. These routines mapped the platform-specific functions to two global MIDIHANDLERS structures, called MIDIDevMIDIHandlers and PCMDevMIDIHandlers, respectively:

```
typedef struct _MIDIhandlers {
  BOOLEAN (*NoteOff)(VOICEOWNER *voiceOwner);
  VOICEOWNER *(*NoteOn)(UINT8 ch, UINT8 note, UINT8 vel,
                        INT8 pan);
  BOOLEAN (*PolyPressure)(UINT8 ch, UINT8 note,
                          UINT8 value);
  BOOLEAN (*Control)(UINT8 ch, UINT8 ctrl, UINT8 value);
  BOOLEAN (*Program)(UINT8 ch, UINT8 program);
  BOOLEAN (*AfterTouch)(UINT8 ch, UINT8 value);
  BOOLEAN (*PitchBend)(VOICEOWNER *voiceOwner,
                       UINT8 vhi, UINT8 vlo);
  BOOLEAN (*SetNoteVelocity)(VOICEOWNER *voiceOwner,
                             UINT8 vel, INT8 pan);
} MIDIHANDLERS;
```

As you can see, MIDIHANDLERS is simply a collection of function pointers. By assigning the platform-specific MIDI device output services at runtime, we support one of the prime directives of our Soundtrack Manager: platform independence. We don't care how a particular platform accomplishes this work, just that it does so. (This assignment happens in \Code\Sound\Win9x\Devices.c, while the actual functions are defined in \Code\Sound\Win9x\MIDIdev.c.) All we need to do to move this library to another platform is to write these two files anew, and put them into a different folder. We retain the file names, function names, arguments and interfaces, but the code within these functions can and will change for different platforms.

Upon successfully initializing each device type, InitDevices sets the MIDIHandlers member of the device structure to the address of the desired MIDIHANDLERS. It is these device- and platform-specific functions that get called from the Soundtrack Managers HandleMIDIFileEventX functions.

MIDI meta-events are handled a bit differently than channel voice messages. We still extract the message-specific data from the `MIDIFILEEVENT`, but there are no platform services involved in their execution. For instance, `HandleMIDIFileEvent-SetTempo` resets the `pulsesPerFrame` member of the `MASTERPLAYER` structure, while `HandleMIDIFileEventEndOfTrack` simply sets the track `PLAYER`'s `isOver` member to `TRUE`.

You may have noticed that some of the `MIDIHANDLERS` function pointers take a `VOICEOWNER` pointer as an argument. This is the structure we use to keep track of all currently sounding MIDI notes. It is a fairly straightforward proposition to use a platform's soundcard or other MIDI hardware device to play back MIDI files. It is quite a different task to provide the actual DA samples for each instrument and note you want to play. Using custom samples frees you from the tyranny and sonic uncertainty of the native sound generation hardware, but involves a great deal more programming effort to realize. We take up this task in Chapter 7 when we build a custom wavetable synthesizer in software, and make full use of our `PCMDevMIDIHandlers`.

MIDI API Redux

In Chapter 5 we presented the Soundtrack Managers public MIDI file API. Let's take another look at how a few of those functions work with our new understanding and appreciation of the `NMD` format.

`MIDIFileSeekToMarker` seeks through the MIDI file to the designated marker number. (If you only know the marker name, you can retrieve the marker number using the `MIDIFileGetMarkerNum` routine.) This function calls our private `SeekMIDIFile` function, passing the pulse of the desired `MIDIFILEMARKER` number. This is turn calls `SeekMIDIFileTrack` on each of the `MASTERPLAYER`'s sub-players. We make use of our `MIDIFILESEEKPOINT`s right away in this routine by doing a coarse seek to the closest seek point before our target pulse. At this seek point pulse, we process any program change `CATCHUPEVENT`s we may have stored, and handle any program changes between the seek point and the desired pulse:

```
for(catchUpEventNum=0;
    catchUpEventNum<seekPoint->nCatchUpEvents;
      catchUpEventNum++) {
  catchUpEvent =
    &(seekPoint->catchUpEvent[catchUpEventNum]);
  if(BuildMIDIFileEvent(&MIDIFileEvent,
      catchUpEvent->event, catchUpEvent->refPulse)) {
    MIDIFileEvent.player = player;
    if((MIDIFileEvent.duration ==
      MIDIFILEEVENT_DURATION_INFINITE) ||
      ((MIDIFileEvent.start + MIDIFileEvent.duration) >
        pulse)) {
      if(MIDIFileEvent.HandleEvent) {
```

```
      if(handleEvents ||
        (MIDIFileEvent.HandleEvent ==
          HandleMIDIFileEventProgram)) {
        MIDIFileEvent.HandleEvent(&MIDIFileEvent);
      }
    }
  }
}
}
//Now process all the events between the seekPoint and
// the desired time offset, if the events don't end
// before that.
track->currentEvent = seekPoint->currentEvent;
track->currentEventNum = seekPoint->currentEventNum;
track->currentPulse = seekPoint->refPulse;
while(BuildMIDIFileEvent(&MIDIFileEvent,
        track->currentEvent, track->currentPulse) &&
          (MIDIFileEvent.start < pulse)) {
  MIDIFileEvent.player = player;
  track->currentPulse = MIDIFileEvent.start;
  if((MIDIFileEvent.duration ==
      MIDIFILEEVENT_DURATION_INFINITE) ||
      ((MIDIFileEvent.start + MIDIFileEvent.duration) >
        pulse)) {
    if(MIDIFileEvent.HandleEvent) {
      if(handleEvents ||
      (MIDIFileEvent.HandleEvent ==
        HandleMIDIFileEventProgram)) {
          MIDIFileEvent.HandleEvent(&MIDIFileEvent);
      }
    }
  }
}
  ...
```

MIDIFileTrackSetMute and MIDIFileSetMute behave very similarly to MIDIFileSeekToMarker. They both seek to the current MASTERPLAYER pulse by calling the private function SeekMIDIFileTrack on one or all of the PLAYERs. This routine is smart enough to turn off any currently sounding notes if we're muting. If we're unmuting, it will process any CATCHUPEVENTs at the seek point and between the seek point and the target pulse. MIDIFileTrackSetMute takes one additional parameter for the type of muting desired, either MIDIFILETRACK_MUTETYPE_INSTANT or MIDIFILETRACK_MUTETYPE_NEXTEVENT. These direct the function to behave as you would expect: muting or unmuting the specified track immediately, or upon encountering the next track event.

The MIDIFileSetMarkerProc function registers a procedure to be called when our NMD parser hits a MIDIFILEMARKER event. The MIDIFILEMARKERPROC takes three arguments: the MASTERPLAYER of the MIDI file, the marker number and the marker name. Having the MASTERPLAYER gives you the powerful ability to call any public MIDIFILE function from within this callback routine. You can also perform a check for a specific marker number or name, and take whatever appropriate action you choose. In Chapter 10, we register a private MIDIFILEMARKERPROC to call ourselves back in our scripting interface.

MIDIFileSetVolume and MIDIFileTrackSetVolume work the same way. The volume and pan fields of either the MASTERPLAYER or PLAYER are set to their new values to affect all upcoming notes. They also adjust the velocity of any currently sounding notes on all or one of the MIDI tracks by calling the SetNoteVelocity member of the particular devices MIDIHANDLERS.

Summary

In this chapter, we presented a case study where the interactive limitations of SMFs were overcome by the development of a new MIDI file format. Designed from the start for musical and interactive use, the new MIDI format, or NMD format, retains the efficiency of the SMF command structure while extending its support for such key behaviors as looping, branching or arbitrary mid-file startup.

7 Building a software synthesizer – gonna build me a robot!

In this chapter, DA samples are used in conjunction with the MIDI services described in Chapter 6 to create a powerful, custom software wavetable synthesizer. The implementation details of the synthesizer are thoroughly presented, and include a robust voice tracking and stealing system.

Strikes one and two

The two main arguments against the use of MIDI in games are: it's not interactive, and it sounds bad. We have overcome the first objection via the new NMD MIDI format presented in Chapter 6. We address the second objection in this chapter.

For years, game machines relied on there being hardware available to generate sounds under MIDI control. Digital audio samples corresponding to the individual notes of the MIDI file were generated in real-time and sent through a digital-to-analog (D/A) converter for amplification. Originally, there was no minimum configuration or set of capabilities you could rely on for any given synthesizer, as they were all different. If you didn't know the exact synthesizer to which you were connected on the other end of the MIDI cable, it was virtually impossible to know what sounds were going to come out. This was a problem.

Enter General MIDI

The first solution to this problem came in the form of the 'General MIDI' (GM) system. Introduced in 1991 by the MMA, the GM system defined a set of capabilities to expect in a compliant synthesizer module. It specified a minimum number of voices, sound and drum note mappings, octave registration, pitch bend range and controller usage. It also defined 128 instrument names, known as the 'GM Sound Set,' that would always be available on any GM system. These names cover a wide range of orchestral, band, percussive and synthetic timbres. However, the MMA did not mandate a specific synthesis technology or level of quality necessary for these sounds. Those critical decisions were left to the individual manufacturers.

As early game machines did not have either the processing or storage resources to support DA sample sets, they instead settled on using synthesis chips to generate their sounds in real time. The most common algorithm to emerge in the GM marketplace for games was FM (Frequency Modulation): the same technology used in radio broadcasting, brought down to the range of human hearing. Early FM chips used a very simple form of this algorithm known as '2-operator' FM, generating all 128 GM sounds from two simple sinusoids. The sounds these chips generated were only rough approximations of the real instrument tones they were meant to represent. Later on, more complex FM algorithms and other synthesis technologies were used to increase the GM sound bank's timbral quality and complexity. With the GM set, you could rely on a given sound name being present in a particular instrument slot, but the sounds themselves varied enormously from one manufacturer to the next. So even with GM, you still couldn't reliably predict how your game was going to sound to the end user. All of this caused a great deal of anguish and posed a real challenge to many a game composer. In spite of these shortcomings, SMFs playing back on GM hardware presented a powerful, easy and efficient way for games to include music on digital audio-challenged systems.

Due to the tremendous increase in computer processing power and cheaper, higher capacity memory chips, audio manufacturers now commonly use DA sample sets exclusively for their MIDI GM sound sets. This is all well and good if your game happens to play back on one of these devices, but it still does not address the variable nature of inter-platform sound sets. We want our games to sound the same no matter what platform we use. So the question is, how can we use DA, and still maintain the interactivity of our NMD MIDI format?

Building a wavetable synthesizer

The answer is: we roll our own wavetable synthesizer in software, of course. Requiring only an ever-present D/A convertor, we free ourselves from the tyranny of unknowable hardware.

The fundamental design of a wavetable synthesizer calls for playing back and controlling custom DA samples, or wavetables, in response to MIDI messages. The sounds themselves can be anything, from a recording of a real instrument to any sound effect. They should further play at a volume controlled by the velocity of the MIDI note on message, and continue to play until a MIDI note off is sent. They should respond appropriately to all supported MIDI controller messages. We should also have the ability to use a single sample for a number of different notes at the user's discretion to save memory. This requires the ability to transpose the original sample's pitch over a specified note range. We also want the ability to layer sounds. This is a technique wherein more than one sound is played for a single MIDI note on message. We must also have a way to cap the number of simultaneously sounding notes or wavetables depending on the resources of the individual game platform. This involves keeping track of what samples are playing on which notes, and stealing those resources as necessary as the music plays.

MIDI response

One of the first of these requirements, that our synthesizer respond to MIDI messages, is relatively easy to achieve. We must simply send all the MIDI messages we receive down to the platform device that handles DA. This is accomplished at the outset of our library in the `InitDevices` routine (in `\Code\Sound\Devices.c`). As we've seen before, this function calls `InitMIDIHandlers` that calls `PCMDevice-MIDIHandlersInit` (in `\Code\Sound\Win9x\DSMix\PCMDev.c`) if `__SLIB_-DEVICE_PCM__` has been defined. This routine sets the function pointers of the global `PCMDevMIDIHandlers` structure to the appropriate private Soundtrack Manager handler functions for the different MIDI events:

```
void PCMDeviceMIDIHandlersInit(void)
{
  PCMDevMIDIHandlers.NoteOff = PCMDeviceNoteOff;
  PCMDevMIDIHandlers.NoteOn = PCMDeviceNoteOn;
  PCMDevMIDIHandlers.SetNoteVelocity = PCMDeviceSetNoteVelocity;
  PCMDevMIDIHandlers.PolyPressure = PCMDevicePolyPressure;
  PCMDevMIDIHandlers.Control = PCMDeviceControl;
  PCMDevMIDIHandlers.Program = PCMDeviceProgram;
  PCMDevMIDIHandlers.AfterTouch = PCMDeviceAfterTouch;
  PCMDevMIDIHandlers.PitchBend = PCMDevicePitchBend;
  return;
}
```

As we'll see in detail below, the bulk of the work happens in these handlers to figure out what sounds to play and how. With the assignment of these handler functions, we're on our way to clearing hurdle number one.

MIDI + DA

A second requirement of our wavetable synthesizer, that DA samples get played in response to MIDI messages, is a bit more complicated. Immediately after `InitMIDI-Handlers` we call `InitVoiceOwners`. This function initializes two global tables inside the Soundtrack Manager we use to keep track of who has what voices at any given time. These tables, called `VoiceOwners` and `VoiceOwnerTable`, require some additional explanation.

As I was walking to St. Ives . . .

We've spoken about notes, keys, sounds and voices. It can be very confusing without a clear understanding of what these terms mean and how they relate to one another. In this book, a "sound" refers to the DA samples themselves, be they previously recorded and stored, or generated in real-time by the hardware. A "voice," on the other hand, is the resource that plays back a "sound." This is typically a specific sample, buffer or DA oscillator on the output platform. It is the thing that voices the sound. A 'key' is one of the white or black bars seen on a conventional piano keyboard or MIDI keyboard controller. A 'note' is the desired musical pitch we want to play and includes both the pitch name and an octave number designation, such as "C2" or "G#4." The MIDI specification assigns each one of these notes a key number for easy binary transmission. The terms "key" and "note" are therefore used interchangeably. In a wavetable synthesizer, the audio artist has to define what sound or sounds to play for a given MIDI key or note number. This is called "key-mapping." The majority of confusion arises when the artist maps more than one sound to a single MIDI note. Since each sound is played by a voice, this means that we could have several voices playing in response to a single MIDI note on message. Put another way, there is a 1-to-1 relationship between a "voice" and a "sound," but there is potentially a 1-to-many relationship between a "note" and the number of "voices" it takes to play that "note."

Mapping sounds

So with all that in mind, the Soundtrack Manager uses a VOICEOWNER structure to keep track of the MIDI channel and note numbers used to play that note, along with the VOICEs holding the SAMPLEs (or sounds) it triggers:

```
typedef struct _voiceowner {
  volatile BOOLEAN   inUse;
  UINT8              channel;
  UINT8              note;
  VOICE              *voice[MAX_SAMPLES_PER_KEY];
  void               (*NoVoicesCallback)(void *);
  void               *callbackParams;
} VOICEOWNER;
```

The first member of this structure is a boolean to indicate whether this VOICEOWNER is currently being used. The next two fields record the MIDI channel and note we're playing. The voice field is an array of VOICE structures that map the note to the specific SAMPLEs it uses:

```
typedef struct _voice {
  char    foo[4];
  SAMPLE  sample;
  UINT32  timeOn;
  UINT32  priority;
  struct _voiceowner *owner;
} VOICE;
```

Up to `MAX_SAMPLES_PER_KEY` SAMPLEs can be mapped to a single MIDI note. This can change depending on the underlying platform, and for this PC version is set to 2. It is important to notice that we keep track of both the time this particular sound was started, and the `VOICE`'s priority. This allows us to perform intelligent note-stealing depending on the number of VOICEs we allow on a particular platform. The final member is a pointer to this `VOICE`'s VOICEOWNER.

The next member of our VOICEOWNER structure is a function pointer to be called when there are no more VOICEs available to play the desired note. In this case, we may steal a VOICEOWNER to play the newer note. When we do this, we call the specified `NoVoicesCallback` function passing the programmer-defined `callbackParams` to do any cleanup or notification before the VOICEOWNER is taken.

As mentioned above, a VOICE holds the SAMPLE for the MIDI note, and these VOICEs are kept in a VOICEOWNER structure for easy bookkeeping. We limit the actual number of simultaneous VOICEs that can play on the platform by setting the `MAX_VOICES` dimension parameter of the global `VoiceTable` array (in `\Code\Sound\Win9X\DSMix\PCMDev.c`).

Notes

There is one more piece of this note and voice puzzle to explore. Successfully calling `InitDevices` brings us to our list of `ModuleProcs` (discussed in Chapter 3). This is a list of function pointers that gets filled in `SM.c` based on the definitions found `modules.h` (both in `\Code\Sound`). If we define `__SLIB_MODULE_MIDIFILE__`, `ModuleProcs` will include a pointer to the `InitMIDIFiles` function. This calls `InitNoteTable` to initialize a static module-wide array called `NoteTable`. This is an array of NOTETABLEENTRYs at the platform-independent level (in `\Code\Sound\MIDIFile.c`):

```
typedef struct _notetableentry
{
  UINT8                   channel;
  UINT8                   note;
  UINT8                   vel;
  INT8                    pan;
  FIXED20_12              timeOff;
  struct _notetableentry  *next;
  VOICEOWNER              *voiceOwner;
} NOTETABLEENTRY;
```

Here we save the MIDI channel, note, velocity, pan and ending time of the given note. We also save the VOICEOWNER returned from the device-type specific MIDI note-on handler in `HandleMIDIFileEventNoteOn`. Each NOTETABLEENTRY is also inserted into the current MIDITRACK's list of `noteOnEvents` inside this track's PLAYER. This is so we can quickly and easily turn off any notes as directed by the score, update all currently sounding notes in response to incoming MIDI messages and process other Soundtrack Manager API commands (such as pause, stop, resume, play, mute or unmute and seek).

Our own API

At this point, we have a pretty good handle on the note- and voice-handling mechanics of our software wavetable synthesizer. But just how do we associate an individual SAMPLE with the MIDI note and channel combination that makes it play? The answer is found in the sample code below:

```
UINT8 chan, lowNote, highNote, centerNote;
UINT16 volume ;
INT16 pan;
BOOLEAN mono;
SAMPLE *sample = NULL;
if(sample = GetSampleFromFile("MyFile.wav", FALSE, FALSE, NULL))
{
  chan = 0;
  lowNote = centerNote = 60; highNote = 68;
  volume = SCALEVOLUME(127, 127);
  pan = SCALEPAN(0, 0);
  mono = TRUE;
  if(MapSampleToRange(sample, chan, lowNote, highNote, centerNote,
                     volume, pan, mono))
  {
    //SUCCESS!
    return(TRUE);
  } else {
    //error mapping SAMPLE
  }
} else {
  //error loading SAMPLE
}
```

This code loads a WAV file into one of our internal SAMPLE structures, and maps the SAMPLE to MIDI channel 0, notes 60 to 68 in MapSampleToRange (in \Code\ Sound\Win9x\Keymap.c). Inside this function, the SAMPLE's centerNote is initially the same as its lowNote. This means we have to transpose this SAMPLE up a maximum of 8 semitones to play all the notes in the specified range. We set the volume and pan of the SAMPLE for the specified note range, and declare this SAMPLE to be a mono sound to our mapping routine.

More on key maps

As mentioned briefly above, a "key map" associates a set of note numbers on a MIDI channel with one or more DA SAMPLEs. To keep track of these key-to-sample channel mappings, the Soundtrack Manager defines a global array of KEYMAP-CHANNELs called SMKeyMap (in \Code\Sound\Win9x\Keymap.c). KeyMapInit initializes this array, called from PCMDeviceInit in the InitDevices routine. A KEYMAPCHANNEL is defined as:

```
typedef struct    {
  INT16           transpose;
  UINT16          bendSensitivity;
  FIXED20_12      bendMult;
  INT16           pan;
  BOOLEAN         sustainOn;
  KEYMAPRANGE     *ranges;
} KEYMAPCHANNEL;
```

The first member of this structure, transpose, defines a base amount of pitch shift for a group of notes and SAMPLEs. All notes within the specified ranges on this MIDI channel (discussed below) will be raised or lowered by the number of semitones in this signed quantity. Transposition is accomplished by multiplying a SAMPLE's sampling frequency by an appropriate scaling factor. If a SAMPLE is to cover a range of notes, we have to multiply the SAMPLE's sampling frequency by a note-specific scaling factor. Any additional MIDI pitch bend messages we receive must be accommodated in this scaling as well. To this end, we store the most-significant-byte of the MIDI pitch bend sensitivity controller message in the bendSensitivity field. This is a special MIDI Registered Parameter Message that defines the total number of semitones covered by the MIDI pitch bend channel voice message. We initialize this field to 12 for a maximum excursion of $+/-$ one octave. The bendMult field stores the scaling factor of an incoming pitch bend message taking this bendSensitivity field into account. Regarding the last two fields, pan sets the left–right position of the SAMPLE, while sustainOn indicates whether the sustain pedal is depressed for this MIDI channel's range of notes.

The final member of KEYMAPCHANNEL is a pointer to a list of KEYMAPRANGE structures that specify the MIDI keys or note numbers to which we want our SAMPLEs to respond:

```
typedef struct _keymaprange {
  UINT8     noteOffset;
  UINT8     nKeys;
  KEYMAPKEY *key;
  struct _keymaprange *next;
} KEYMAPRANGE;
```

The lowest note of the mapping is stored in `noteOffset`, while `nKeys` stores the total number of keys over which the current `SAMPLE` is used. The `key` field is a pointer to an array of `KEYMAPKEY` structures that tell us what sounds are mapped to each of the `nKeys` in this `KEYMAPRANGE`:

```
typedef struct {
  KEYMAPSAMPLEINFO  sampleInfo[MAX_SAMPLES_PER_KEY];
  BOOLEAN           sustainOn;
  BOOLEAN           on;
} KEYMAPKEY;
```

These are where we reference the actual `SAMPLE`s used for each `KEYMAPKEY` of the given `KEYMAPRANGE`. `sampleInfo` is an array of `MAX_SAMPLES_PER_KEY` `KEYMAPSAMPLEINFO` structures:

```
typedef struct _keymapsampleinfo {
  SAMPLE   *sample;
  UINT16   volume;
  INT16    pan;
  BOOLEAN  isMono;
  UINT8    centerNote;
} KEYMAPSAMPLEINFO;
```

`sample` points to a `SAMPLE` structure that holds the DA data itself and its associated information (see Chapter 4). We may use the given `SAMPLE` in a number of different places within the Soundtrack Manager, and want to be able to specify the volume and pan of each individually. For this reason, each `KEYMAPSAMPLEINFO` has its own `volume` and `pan` member variables. We flag whether the `SAMPLE` is a mono sound in the `isMono` field, and set the center note of the `SAMPLE` in `centerNote`. This last field defines the MIDI note number where the `SAMPLE` is played back without any transposition.

Popping the stack of our key map explanation back into the `KEYMAPKEY` structure, we use the `sustainOn` field to keep a local copy of the `KEYMAPCHANNEL` `sustainOn` field. This gets set or cleared by the MIDI sustain pedal controller and either disables or enables all note off messages on the given MIDI channel. (We'll discuss the significance of this behavior at the end of this chapter.) The final `on` field is set to `TRUE` when we receive a MIDI note on message for this key in our channel's key map, and `FALSE` when we receive the corresponding note off message.

One final comment regarding these two fields: when the sustain pedal is pressed, all keys set to `on` will set their `sustainOn` members to `TRUE` as well. This means those keys' `SAMPLE`s will continue to play even if a note off message comes along. When the sustain pedal is released, any keys that received note off messages but are still playing because of their sustain status will be turned off in `UpdateSamples`.

Popping our explanation stack once more returns us to the final field of the KEYMAPRANGE structure. The next field reveals that it is possible to specify any number of different key ranges for a given set of sounds.

Sounds and keys

Now that you've seen all of the structures we use to support sample-to-key mapping, let's go through the MapSampleToRange code in greater detail. Please refer to the code listing in \Code\Sound\Win9x\Keymap.c.

After checking the validity of the input channel, low note, high note and center note parameters, we see if there is an existing KEYMAPRANGE on this MIDI channel that covers the input note range. If there is, we may have to extend it to accommodate our new input range:

```
highNote = MAX(lowNote, highNote);
if(range=SMKeyMap[channel].ranges) {
  //A key range already exists; have to extend?
  if((range->noteOffset > lowNote) ||
    ((range->noteOffset + range->nKeys-1) < highNote)) {
    //Must extend key range by allocating a bigger one
    // and copying the old one into it
    if(newRange = (KEYMAPRANGE*)SMGetMem(sizeof(KEYMAPRANGE))) {
        newLowNote = MIN(lowNote, range->noteOffset);
        newHighNote = MAX(highNote,
                        (range->noteOffset + range->nKeys-1));
        newRange->noteOffset = newLowNote;
        newRange->nKeys = newHighNote - newLowNote + 1;
        newRange->next = NULL;
        if(newRange->key = (KEYMAPKEY*)SMGetMem(newRange->nKeys *
          sizeof(KEYMAPKEY)))                              {
        //Initialize range info
        APPMEMSET(newRange->key, 0, newRange->nKeys * sizeof(KEYMAP-
          KEY));
        //Copy old key info to new key range
        memcpy(&(newRange->key[range->noteOffset -
              newRange->noteOffset]), range->key,
        (range->nKeys * sizeof(KEYMAPKEY)));
    SMFreeMem(range->key);
    SMFreeMem(range);
    range = newRange;
    SMKeyMap[channel].ranges = range;
    result = TRUE;
  } ...
```

If we do not find a corresponding `KEYMAPRANGE` for this, we allocate a new one, including space for all the keys:

```
if(range = (KEYMAPRANGE*)SMGetMem(sizeof(KEYMAPRANGE)))
{
  range->noteOffset = lowNote;
  range->nKeys = highNote - lowNote + 1;
  range->next = NULL;
  if(range->key =
  (KEYMAPKEY*)SMGetMem(range->nKeys*sizeof(KEYMAPKEY)))
  {
    APPMEMSET(range->key, 0, range->nKeys*sizeof(KEYMAPKEY));
    SMKeyMap[channel].ranges = range;
    result = TRUE;
  } else {
    SMFreeMem(range);
  }
}
```

Once we have a valid key range, big enough to hold all the keys, we map the `SAMPLE` to each key in the range by filling the key's `sampleInfo` structure:

```
for(note=lowNote; note<=highNote; note++)
{
  key = &(range->key[note - range->noteOffset]);
  // find the next blank sample slot
  for(i=0; (i<MAX_SAMPLES_PER_KEY) &&
              (key->sampleInfo[i].sample); i++) ;
  if(i<MAX_SAMPLES_PER_KEY)
  {
    key->sampleInfo[i].sample = sample;
    key->sampleInfo[i].volume = volume;
    key->sampleInfo[i].pan = pan;
    key->sampleInfo[i].isMono = mono;
    key->sampleInfo[i].centerNote = centerNote;
  } else {
    //error - tried to map too many samples to this note
  }
}
```

Congratulations! You now possess a greater understanding of what it means to map a `SAMPLE` to a particular MIDI channel and key range, and how we accomplish this specifically inside our `MapSampleToRange` routine. Now let's put that knowledge to use and finish making the wavetable synthesizer.

Playing MIDI controlled samples

Back in `MIDIFile.c`, we call the `HandleMIDIFileEventNoteOn` function to process any MIDI note on events. This takes a `MIDIFILEEVENT` as input, and extracts the channel, note and velocity parameters from the contained `MIDIMESSAGE`. The velocity and pan undergo several adjustments for the MIDI file, track and master music settings. If the event's MIDI channel is mapped to a valid device, we get a free `NoteTable` entry and call the specific devices note on MIDI handler for this channel:

```
voiceOwner = Devices[MIDIMap[channel]].MIDIHandlers->
             NoteOn(channel, note, vel, pan);
```

This is a rather long-winded way to call a function. We do it this way because we: (1) allow different MIDI channels to use different devices, and (2) support multiple platforms through the use of MIDI handler function pointers. This MIDI channel-to-device mapping happens in `MIDIMapsInit` (in `\Code\Sound\Win9x\Devices.c`). This routine calls `InitializeMIDIMap` to set the device used for each channel, and sets a global `MIDIMap` variable to keep track of those assignments. By defining `__SLIB_MODULE_SAMPLE__` and `__SLIB_DEVICE_PCM__` in `modules.h` and setting the `MIDIMap` to use the PCM device for all channels, we direct the Soundtrack Manager to use the global `PCMDevMIDIHandlers` functions. In the code above, this means `PCMDeviceNoteOn` will be called to handle this note on message (in `\Code\Sound\Win9x\DSMix\Pcmdev.c`).

Inside `PCMDeviceNoteOn`, the first thing we do is call `KeyMapGetKey` to return the `KEYMAPKEY` for this channel and note combination:

```
KEYMAPKEY* KeyMapGetKey(UINT8 channel, UINT8 note) {
  KEYMAPRANGE *range;
  KEYMAPKEY *key = NULL;
  if(range=SMKeyMap[channel].ranges) {
    if((note >= range->noteOffset) &&
       (note < range->noteOffset + range->nKeys)) {
      key = &(range->key[note - range->noteOffset]);
    }
  }
  return(key);
}
```

We next get a free `VoiceOwnerTable` entry by calling, oddly enough, `GetFree-VoiceOwnerTableEntry`. This routine returns the first available `VOICEOWNER` in that table. For each of the `MAX_SAMPLES_PER_KEY` of this key, we call

`NoteGetSample` to find the `SAMPLE` mapped to this key:

```
KEYMAPSAMPLEINFO* NoteGetSample(UINT8 channel, UINT8 note,
UINT32 sampleIndex)
{
  KEYMAPRANGE *range;
  KEYMAPKEY *key;
  KEYMAPSAMPLEINFO *sampleInfo = NULL;
  if(range=SMKeyMap[channel].ranges) {
    if((note >= range->noteOffset) &&
        (note < range->noteOffset = range->nKeys)) {
      key = &(range->key[note - range->noteOffset]);
      if(key->sampleInfo[sampleIndex].sample) {
        sampleInfo = &key->sampleInfo[sampleIndex];
      }
    }
  }
  return(sampleInfo);
}
```

If there is such a `SAMPLE`, we pass it on to `GetVoice` (in `\Code\Sound\Win9x\ DSMix\Pcmdev.c`). This is where we perform all our platform-specific voice alloca-tion and stealing. A `VOICE` is the structure we use to hold an individual playing `SAMPLE` in our synthesizer:

```
typedef struct _voice {
  char      foo[4];
  SAMPLE    sample;
  UINT32    timeOn;
  UINT32    priority;
  struct    _voiceowner *owner;
} VOICE;
```

We keep a pointer to the `SAMPLE` in `sample`, save its `timeOn`, `priority` and `owner`, and keep a local static table of `MAX_VOICES` `VOICE`s in `VoiceTable`. Inside `GetVoice`, if the `SAMPLE` we want to play is already playing, we stop that `SAMPLE` and re-use that same `VOICE`. Otherwise, we have to search the `VoiceTable` using different criteria. On the PC, if the `SAMPLE`'s DirectSound buffer is `NULL`, that `VOICE` is free. If we still haven't found a free `VOICE`, we keep track of the oldest playing `VOICE`, while trying to find a `VOICE` that's done playing but still has a DirectSound buffer. We steal this `VOICE` first, and steal the oldest voice as a last resort.

Once we select a `VOICE` to use for this `SAMPLE`, we duplicate the input `SAMPLE`'s DirectSound buffer into the free `VOICE`'s `SAMPLE`, and return the selected `VOICE`. The reason we call `DuplicateSoundBuffer` within `GetVoice` is twofold: it saves space, and we can change the parameters of each buffer independently. This creates a new secondary buffer that shares the original buffer's memory, but allows each to be played, changed or stopped without affecting the other.

Another place where the Soundtrack Manager can be made to support additional platforms is by rewriting this `PCMDev.c` file. Preserving the interface by retaining all the function names and argument lists, we only change the internal instructions of the subroutines themselves. This new file is stored in a different folder than the PC version, and included in a new project for the platform at hand.

The well-tempered synth

Once we have our `VOICE`, we have to set the frequency of its `SAMPLE` to match the desired note value. This involves multiplying the `SAMPLE`'s sampling frequency by the correct amount to achieve the desired transposition. Rather than calculating a semitone multiplier for every MIDI note in the score, we keep a static table of pre-calculated, equal-tempered semitone sampling frequency scaling factors in the static `EqualTemp` array (in `\Code\Sound\Win9x\DSMix\PCMDev.c`):

```
static FIXED4_12 EqualTemp[] = {
  ...
  0x0800, 0x0879, 0x08FA, 0x0983,
  0x0A14, 0x0AAD, 0x0B50, 0x0BFC,
  0x0CB2, 0x0D74, 0x0E41, 0x0F1A,

  0x1000, 0x10F3, 0x11F5, 0x1306,
  0x1428, 0x155B, 0x16A0, 0x17F9,
  0x1965, 0x1AE8, 0x1C82, 0x1E34,
  ...
  0x8000
};
```

Throughout the Soundtrack Manager, we make use of 16- and 32-bit fractional integers to do multiplication. This avoids the use of expensive or nonexistent floating-point calculations, depending on the platform, while preserving the resolution floating-point operations provide. In this case, we define `EqualTemp` to be a `FIXED4_12` array. This is simply a re-named UINT16, with 4-bits of whole number and 12-bits of fraction. We can also think of this as multiplying by 4096, or 0x1000 in hexadecimal. For example, an octave is a doubling of the frequency, so 4096 * 2 = 8192, or 0x2000. In Western music, an octave is divided into 12 equal semitones, or half-steps, known as the "equal-tempered" tuning system. Each semitone is a constant proportion of the note immediately adjacent to it, rather than a fixed number of Hertz. This proportion is equal to the 12th root of 2. To find the frequency of the sound a 1/2-step higher, you simply multiply the current frequency by 1.05946. Two 1/2-steps equals one whole step = $(1.05946)^2$, or 1.12246. Keep doing this 12 times, and you reach 2.0, or a doubling of the frequency that is the octave. We store a semitone's scaling factor as 4096 * 1.05946 = 4339.54816. Truncating this quantity to a whole number and converting to hex yields 0x10F3. To find the frequency of a sound a 1/2-step lower, we multiply by the inverse of that semitone proportion, or approximately 0.94387, yielding (4096 * 0.94387) = 3866.09152 = 0x0F1A. The `EqualTemp` array, therefore, is merely a table of +/− 3 octaves' worth of scaling factors in `FIXED4_12` format.

We index this table upon each MIDI note on message using a combination of the MIDI note number, the `SAMPLE`'s `centerNote`, and the `KEYMAPCHANNEL`'s `transpose` value, and multiply the `SAMPLE`'s sampling frequency by this stored factor:

```
#define LIMIT(kl, x, kh) MIN(MAX((kl),(x)),(kh))
FIXED20_12 freqMult; // 32-bit integer, 12-bit fraction
DWORD freq;
freqMult = (FIXED20_12)EqualTemp[LIMIT(0,
  36+(note-sampleInfo->centerNote)+SMKeyMap[ch].transpose,
  72)];
freq =
  (DWORD)(((FIXED20_12)(sample->freq)*freqMult)>>12);
```

Any pitch bend information on this channel is also taken into account by multiplying that new sampling frequency by the `KEYMAPCHANNEL` `bendMult` scalar (the right-shift 12 being equivalent to dividing by 4096):

```
if(SMKeyMap[ch].bendSensitivity > 0) {
  freq = (freq * SMKeyMap[ch].bendMult)>>12;
}
```

We set the DirectSound buffer sampling frequency to this quantity when we call `SampleSetFreq`.

We finally make some noise

The `SAMPLE` is finally played in `PCMDeviceNoteOn` by calling `PlaySample`. The `VOICEOWNER`'s `VOICE` is set to the `VOICE` we found earlier, while the owner member of the `VOICE` structure points back to this `VOICEOWNER`. If the sustain pedal is on, we set the `sustainOn` field of the `KEYMAPKEY` we got at the head of this routine. The `channel`, `note` and `inUse` fields of the `VOICEOWNER` are set, and the key is set to `on`.

Other coolness

There are some other cool things we can do to and with `SAMPLE`s in our synthesizer that relate directly to its perceived sound quality. The first of these has to do with a `SAMPLE`'s amplitude envelope. The volume of all naturally occurring sounds changes over time. Some sounds rise very quickly to their maximum volume at the outset and gradually fade away. Examples of this kind of sound include piano tones, drum hits and speech. Other sounds take more time to rise to their maximum volume, but cut off very abruptly. A reed organ is a one example of this kind of sound. It turns out that our perception of a sound is greatly influenced by a sound's temporal amplitude changes. Anyone who's ever heard music played backward knows this to be true. The overall frequency content of a reversed sound is the same as the forward version, but our perception of it is completely different.

Envelopes

In synthesis, a sound's amplitude changes through time are characterized by an attack, decay, sustain and release envelope, or ADSR. The 'attack' (A) portion of a sound describes how quickly a sound rises from silence to its maximum volume. The 'decay' (D) portion represents the time it takes a sound to fall back from its maximum volume to a steady-state level. The 'sustain' (S) portion is the time it spends at that level, while the 'release' (R) segment expresses the time it takes for the sound to fade from its sustain level back down to silence.

The Soundtrack Manager wavetable synthesizer accommodates these perceptually significant volume characteristics by directly manipulating the attack and release times of a SAMPLE via the envelope field inside its SAMPLEMIDIINFO structure:

```
typedef struct _samplemidiinfo {
  UINT8             currentNote;
  BOOLEAN           isDrum;
  struct _sample_t  *mutEx;
  SAMPLEENVELOPE    envelope;
  struct _voice     *monoVoice;
} SAMPLEMIDIINFO;

typedef struct _envelope_t {
  UINT32    attackTime;
  UINT32    releaseTime;
} envelope_t, SAMPLEENVELOPE;
```

The SampleSetEnvelope function sets the internal attackTime and releaseTime of a sound's SAMPLEENVELOPE. When we receive a MIDI note on message for a key-mapped SAMPLE, we set the volume of the SAMPLE to 0 by calling SampleSetVolume from within PCMDeviceNoteOn (in \Code\Sound\Win9x\ DSMix\PCMDev.c):

```
if(sample->MIDI.envelope.attackTime) {
  SampleSetVolume(sample, 0, 0);
  SampleSetFade(sample, vel, pan,
    sample->MIDI.envelope.attackTime, FALSE);
}
```

Calling `SampleSetFade` indicates we want to ramp up to the `SAMPLE`'s to desired maximum volume and pan position over `attackTime` milliseconds. This routine calculates the volume change per timer interrupt necessary to accomplish the fade over the specified time, and fills out the `SAMPLE`'s `FADEINFO` structure. All `SAMPLE` fades are handled inside the `UpdateSamples` routine that calls `SampleSetVolume` on every timer interrupt to adjust the `SAMPLE`'s current volume. The periodic timer interval, therefore, determines a fade's fastest speed. In the case of the PC, this equals 8 ms.

A similar process occurs in reverse when we receive a MIDI note off message. We call `SampleSetFade` with a target volume of 0 and a fade time of `releaseTime` milliseconds, and let `UpdateSamples` do its thing.

Drum line

Most musical instrument sounds have a definite end time as well as a start time. There are many instrument sounds that do not exhibit a well-defined cutoff, however. Drums and other percussion instruments are generally left to resonate and fade away of their own accord. Still other percussion sounds are mutually exclusive, meaning one sound interrupts another as in the case of an open versus closed hi-hat cymbal. We accommodate this kind of instrument behavior in our wavetable synthesizer. The `SampleSetDrum` function sets the state of the `SAMPLEMIDIINFO` `isDrum` member. If set to `TRUE`, we ignore all note off messages for this `SAMPLE` and release it in `UpdateSamples` when it's done playing. We use the `SAMPLEMIDIINFO` `mutEx` field to identify another `SAMPLE` whose playback should be terminated when this one is played:

```
if(PlaySample(sample)) {
  sample->MIDI.currentNote = note;
  if(sample->MIDI.mutEx) {
    StopSample(sample->MIDI.mutEx);
  }
  voiceOwner->voice[i] = voice;
  voice->owner = voiceOwner;
}
```

It's not just for music anymore

Finally, while our software wavetable synthesizer was designed to play back instrument samples under MIDI control for musical purposes, there is every reason to trigger DA sound effects with MIDI notes. As mentioned above, a `SAMPLE` can contain any sound. Mapping a sound effect to any MIDI channel and key combination is the same as mapping an instrument sound. At this point in our discussion, these notes have to be included in an NMD file for those sounds to be heard. Later on in Chapter 10, we'll show how to trigger these sounds individually without the need for a MIDI file when we discuss the Soundtrack Manager's English-language script interface.

Summary

In this chapter, we presented the motivations behind and the creation and implementation of a custom wavetable software synthesizer. With this synthesizer, the Soundtrack Manager provides another powerful tool for artistic and interactive audio expression across all platforms. Game audio artists gain a compelling resource for their musical bag of tricks, and end users get a high quality and consistent audio experience.

8 Case study 2: Improving audio performance – faster, you fool, faster!

Processor usage is always an issue in video games. Games often have to accommodate a wide range of machine performance, and the audio engine can't slow down the video frame rate. This chapter details the implementation of a low-level, assembly-language audio mixer developed to improve the CPU performance of the audio engine on the PC, without violating the platform-independence of the higher-level services.

Making a little bit go farther

As mentioned in many places throughout this book, audio gets only about 10% of a machine's processing resources. PC, handheld or console, its always the same story. The audio programming challenge is to find a way to do more in the same amount of time. In the last chapter, we implemented a software wavetable synthesizer on the PC, and used the DirectSound API to do all the mixing. That API, while certainly adequate, is a general-purpose DA solution. It wasn't designed specifically for doing wavetable synthesis, and therefore suffers from some unnecessary overhead when performing this task. If we can eliminate that overhead and do the mixing ourselves, we can make our 10% of the machine go further.

In this chapter, we bypass the DirectSound API and detail the implementation of our own audio mixer. The mixing procedure itself is written in floating-point assembly language to make the most out of our CPU-cycle budget. This frees us up to do some additional signal processing we didn't have the time or capability to do before. This includes reverberation, flanging, ping-pong (forward-then-backward) looping and sample-rate conversion for pitch-shifting. We perform all the MIDI parsing as before, but now directly handle all the samples, instrument changes, volume control and additional effects ourselves. All DA, both MIDI-controlled and otherwise, is mixed internally. We use DirectSound solely as a DA output pipe by placing the final mix into a single DirectSound secondary buffer.

How do we do it?

In this new version of our library, we define a handful of new effects parameters and extend the DA interface to include a number of additional VOICE manipulations. Following the Soundtrack Manager's compartmentalized architecture, we place all the affected files in a new directory called \Code\Sound\Win9x\Win9xMix. This includes Sample.c and all of our DA format files (WAV, AIF, ADPCM, RAM and RAW). The new floating-point mixer code is found in the Voicea.asm file in the FP folder in that directory, along with new versions of the Devices.h and PCMDev.c files. While the underlying DA operations have changed significantly in this new version of our library, the API has not. Due to the Soundtrack Manager's design, all the high-level interfaces we've developed and discussed so far remain unchanged (but for some minor changes to the public PlaySample and StopSample routines which we'll talk about later on).

The new mixer function prototypes, listed near the end of Devices.h, are:

```
void VoiceMix8Mono(UINT8* bufDest, UINT32 nBytes);
void VoiceMix8Stereo(UINT8* bufDest, UINT32 nBytes);
void VoiceMix16Mono(UINT8* bufDest, UINT32 nBytes);
void VoiceMix16Stereo(UINT8* bufDest, UINT32 nBytes);
```

As their names imply, these four routines generate either 8- or 16-bit output data in monaural or stereo format. Input SAMPLE data that does not conform to the chosen mixer output format is converted on the fly. We typically use the VoiceMix16-Stereo version of the mixer, and run the library at a sampling rate of 22 050 Hz. However, the sampling rate of the mixer can be set to any frequency. We adjust the rate at which we step through each SAMPLE individually taking into account its own sampling frequency, the selected output sampling frequency of the Soundtrack Manager and the amount of transposition or pitch-shifting desired. We'll see a bit later on how we select the right mixer and tap into the services of these assembly routines.

Structural changes

A number of structures have changed in this version of the Soundtrack Manager. The PCMdevice structure still contains pointers to the DirectSound object and primary buffer, and retains its boolean status flags. We lose the linked list of DirectSound secondary buffers, and add two new members, DSMixBuffer and MixProc:

```
typedef struct {
  LPDIRECTSOUND lpDirectSound;
  LPDIRECTSOUNDBUFFER lpDSPrimaryBuffer;
  DSMIXBUFFERINFO DSMixBuffer;
  void (*MixProc)(UINT8* bufDest, UINT32 nBytes);
  BOOLEAN bDSInitialized;
  BOOLEAN bDSActive;
} PCMdevice;
```

`MixProc` is a function pointer to which we assign one of the four mixing routines above at runtime. `DSMixBuffer` is a `DSMIXBUFFERINFO` structure that contains a pointer to the single DirectSound secondary buffer into which we place our mix, as well as some additional size and indexing information:

```
typedef struct {
  LPDIRECTSOUNDBUFER lpDSB;
  UINT32 bufferSize;
  UINT32 bufferSegSize;
  UINT32 timerSegSize;
  UINT32 nextWriteOffset;
  UINT32 lastUpdatePos;
  INTERMEDIATEBUFFERINFO intermediateBuffer;
} DSMIXBUFFERINFO;
```

Division of labor

The DirectSound secondary `bufferSize` is divided into `PCMDEVICE_MIXBUFFER_-NUMSEGMENTS` segments. Each `bufferSegSize` is large enough to hold `PCMDEVICE_MIXBUFFER_MSECS_PER_SEGMENT` milliseconds worth of data. These values are set in the `PCMDeviceSetOutputFormat` function called from `PCMDevice-Activate` (both in `\Code\Sound\Win9x\Win9xMix\FP\PCMdev.c`) and defined in `Devices.h`. `timerSegSize` is the number of bytes we can play between each timer interrupt. We keep track of this value because there may be occasions when the interrupt timer is delayed or we've not gotten timely updates of the play cursor from DirectSound indicating how much data has been played out since last time we checked. In either of these cases, we continue mixing `timerSegSize` chunks until the amount left to mix in this frame is smaller than this number. `nextWriteOffset` saves the offset within the DirectSound buffer to fill next, while `lastUpdatePos` saves the buffer offset of our last update.

The last member of the `DSMIXBUFFERINFO` structure is `intermediateBuffer`, which looks like this:

```
typedef struct {
  UINT8* data;
  UINT32 size;
  UINT32 curFillPos;
  UINT32 curPlayPos;
} INTERMEDIATEBUFFERINFO;
```

`data` points to a local buffer into which we do our mixing rather than using the DirectSound secondary buffer directly. This avoids collisions with DirectSound when we have to access and lock its buffer for our mixing purposes. `size` gets set in `PCMDevice-SetOutputFormat` and is twice the size of `DSMixBuffer.bufferSegSize`. We simply mix into our own buffer, using `curFillPos` and `curPlayPos`, and copy from `data` into the DirectSound buffer.

SAMPLEs grow up

Back in Chapter 4, we described the `SAMPLE` structure we use to encapsulate DA in the DirectSound mixer version of the Soundtrack Manager. This structure has been re-arranged for the custom, floating-point mixer of this chapter, and looks like this (see `\Code\Sound\Win9X\Win9xMix\Sample.h`):

```
typedef struct _sample_t {
  UINT8            *data;
  SAMPLEFILEINFO    file;
  SAMPLEPLAYINFO    play;
  SAMPLESTREAMINFO  stream;
  FADEINFO          fade;
  SAMPLELOOPINFO    loop;
  SAMPLEMIDIINFO    MIDI;
  struct _sample_t *next;
} SAMPLE;
```

We've consolidated a lot of the information in the original `SAMPLE` structure into more meaningful substructures. The `FADEINFO` structure is unchanged from before, but everything else is different. We've retained the original calling interface, but changed the underlying behavior to support new functionality. This library's client doesn't see these differences, as they are encased within the private `__SLIB__` portions of our code. Again, it is possible for the game or multimedia application to see these private areas, but they would deliberately have to violate this pre-processor definition's encapsulation to do so. Plus, unpredictable results could ensue should they try to use more than the public interface.

We have talked at length about moving the Soundtrack Manager to different platforms. It is also important to note that we can accommodate changes within a platform as well. The architecture of the Soundtrack Manager insulates the high- from the low-level portions of the code. So from the user's perspective, the impact of moving to a different platform or changing the way we do things on the current platform is the same: zero.

Popping the lid

Well, under the hood it's not zero, so let's take a deeper look at our new `SAMPLE`. `data` is a pointer to the DA samples themselves in RAM. Whether loaded all at once or streamed, we store all the information about the file whence that data came in the `SAMPLEFILEINFO file` member:

```
typedef struct _samplefileinfo {
    char          *name;
    FILEHANDLE    hFile;
    UINT32        freq;
    UINT32        bitWidth;
    UINT32        nChannels;
    UINT32        length;
    UINT32        size;
    UINT32        dataOffset;
    UINT32        curFillPosition;
    void          *compressionInfo;
    BOOLEAN       (*Open)(void *sample);
    BOOLEAN       (*Reset)(void *sample, UINT32 fileDataOffset,
                           UINT32  dataOffset);
    BOOLEAN       (*Read)(void *sample, UINT32 length,
                          UINT8*  data, UINT32* uChkErr);
    void          (*Close)(void *sample);
} SAMPLEFILEINFO;
```

We store the name of the file in `name` so that we can easily reload it upon activation, and `hFile` is the file handle returned when we open the file. `freq`, `bitWidth` and `nChannels` store the sampling frequency, bits per sample and number of channels (one or two) of this `SAMPLE`. The total number of data bytes is kept in `size`, while `length` represents the total number of sample frames. A sample frame is the number of bytes needed for a single sample times the number of channels. `dataOffset` stores the offset to the first sample of DA within the file, while `curFillPosition` keeps track of how far into the file we've read. This helps to speed up file restarts and resets by avoiding a file seek operation unless we absolutely have to do it. The `compressionInfo` member is a structure used to hold any specific information needed to decompress the `SAMPLE` data. This is defined to be a `void` pointer to accommodate any compression scheme. For instance, when decoding ADPCM data, this field contains a pointer to a `SAMPLEADPCMINFO` structure. To add MP3 support, you could define a `SAMPLEMP3INFO` structure and use that to keep track of the decoding parameters between frames.

The last four members of the `SAMPLEFILEINFO` structure are function pointers that are assigned by format-specific parsing. This has to do directly with the `SAMPLE` submodule formats defined in `modules.h`. If `__SLIB_SUBMODULE_WAV__` is defined and we detect a WAV file, the `Reset` function pointer gets set to `WAVResetFile` and the `Read` pointer to `WAVReadFile` (in `\Code\Sound\Win9x\Win9xMix\Wav.c`) All this happens within `SampleParseFileType` which is called from `GetSample-FromFile` (both in `\Code\Sound\Win9x\Win9xMix\Sample.c`). If `__SLIB_-SUBMODULE_ADPCM__` is defined and we detect an ADPCM file, `Read` gets set to

ADPCMReadFile while Reset is set to ADPCMResetFile. The Open and Close function pointers are only assigned if there is some additional processing we want to do on the SAMPLE after we open and before we close the file, respectively. Assigning these function pointers at runtime gives us the ability to use and parse multiple file formats simultaneously, all through the same SAMPLE interface.

The SAMPLEPLAYINFO play member follows next in this new SAMPLE structure. As its name implies, this element contains all the information we use while playing the SAMPLE:

```
typedef struct _sampleplayinfo {
  UINT16 volume;
  INT16 pan;
  UINT32 freq;
  UINT32 type;
  struct _sample_t *playNext;
  void (*DonePlayingCB)(struct _sample_t *sample);
  struct _voice *voice;
} SAMPLEPLAYINFO;
```

The volume, pan and freq sampling frequency data members from our original SAMPLE structure find a home here, as does the playNext SAMPLE pointer and DonePlayingCB function pointer. Next comes a new element, called type, which can take on one of three values to control a SAMPLE's playback behavior. When set to SAMPLE_PLAY_NEW_INSTANCE, type directs the Soundtrack Manager to play multiple copies of the same sound simultaneously. Upon each play command, a new instance of the SAMPLE will be started. This is very useful for short, repetitive sounds like gunshots or footsteps. The sounds can overlap, and each instance of the SAMPLE plays out completely. This is the default SAMPLE type for sounds that have been fully loaded into memory. Because of the difficulty involved in pulling data off a disk from different file locations simultaneously, the SAMPLE_PLAY_NEW_INSTANCE is not permitted for streaming sounds.

The second new SAMPLE type, called SAMPLE_PLAY_RESTART, allows only one instance of the sound to be played at a time. If the sound hasn't finished when another 'play' command is issued for this SAMPLE, the sound will immediately interrupt itself, rewind to the beginning and start playing again. This is the default type for all streaming SAMPLEs. You can create some very cool hip-hop and stuttering effects using this type.

The final new SAMPLE type is called SAMPLE_PLAY_UNINTERRUPTABLE. As the name implies, the sound cannot be interrupted by a new 'play' command to the same SAMPLE if it hasn't finished playing. Both streaming and memory-resident SAMPLEs can be set to this type.

The last member of the SAMPLEPLAYINFO structure is voice. This is a pointer to one of the VOICEs in the global VoiceTable array. This is returned by GetVoice (in \Code\Sound\Win9x\Win9xMix\FP\PCMDev.c) when we call PlaySample. GetVoice sets the isPlaying member of the free VOICE to TRUE, signaling the custom mixer to process this VOICE's SAMPLE. On every timer interrupt, Sound Manager calls UpdateDevices which calls PCMDeviceUpdate. This is a work-horse routine that calculates all the relevant buffer positions for reading and writing new audio data. When we're ready to mix more data, we extract the pointer to the PCMdevice from the global Devices array and call MixProc to mix all playing VOICEs.

The next member of the new SAMPLE structure is a SAMPLESTREAMINFO structure called stream. This holds all information relevant to a streaming SAMPLE, and looks like this:

```
typedef struct _streamdata {
  BOOLEAN    isStreaming;
  UINT32     bufferNSamples;
  UINT32     bufferSize;
  UINT32     bufferSegSize;
  UINT32     segOffset[2];
  UINT32     nextWriteOffset;
  UINT32     jumpToOffset;
  UINT32     lastUpdatePos;
  BOOLEAN    foundEnd;
  UINT32     lengthToStop;
  BOOLEAN    isReset;
} SAMPLESTREAMINFO;
```

The first several members of this structure are calculated and set in the GetSampleFromFile function (in \Code\Sound\Win9x\Win9xMix\Sample.c) when stream is TRUE:

```
sample->play.type = SAMPLE_PLAY_RESTART;
sample->stream.isStreaming = TRUE;
sample->stream.bufferSegSize =
  ((UINT32)(sample->file.freq *
           sample->file.bitWidth/8 *
           sample->file.nChannels) *
  (SAMPLE_STREAMBUFFERSEGMSECS / 10) / 100);
sample->stream.bufferSize =
  sample->stream.bufferSegSize * 2;
sample->stream.bufferNSamples =
  sample->stream.bufferSize /
  ((sample->file.bitWidth/8)*sample->file.nChannels);
if(sample->data = SMGetMem(sample->stream.bufferSize)) {
  if(SampleResetBuffer(sample)) {
    result = TRUE;
  }
}
```

The SAMPLE's play type field is set to SAMPLE_PLAY_RESTART, and the stream isStreaming member is set to TRUE. Using the SAMPLE's file attributes, we calculate how many bytes it takes to hold SAMPLE_STREAMBUFFERSEGMSECS milliseconds worth of data (defined in \Code\Sound\Win9x\Win9xMix\Sample.h), and store this value in bufferSegSize. We need two buffer segments to perform our streaming operations, one to fill and one to play. We therefore set bufferSize to be twice the size of bufferSegSize. The size of the buffer in sample frames is calculated and stored in bufferNSamples. We allocate the memory we'll need to stream this SAMPLE, store that pointer in the SAMPLE's data member and reset the SAMPLE to point to the beginning of the file to be streamed.

The next several data members of the SAMPLESTREAMINFO structure are manipulated and set in UpdateSamples (in \Code\Sound\Win9x\Win9xMix\Sample.c). segOffset is a two-element array that holds the current file offsets for the start of each streaming buffer segment the next time they need data. nextWriteOffset keeps track of where we need to write the mixer's output data, while lastUpdatePos stores the buffer offset of the last update. This is used in conjunction with the VOICE's current position to calculate how far the playback has progressed since the last timer interrupt. We can also specify a sample offset to jump to upon the next update in jumpToOffset. lengthToStop tells how many more bytes there are to play in this stream, and foundEnd is set by SampleFillBuffer when it gets to the end of a sample file's data. The final data member of this stream, isReset, is a flag to speed up multiple consecutive resets without doing any file seeks.

A FADEINFO structure follows next. This was discussed in Chapter 4, and remains unchanged. This is because it's not a SAMPLE-specific structure but a volume-handling construct used throughout the library.

Loopitude

In Chapter 4 we also talked about looping SAMPLEs. In that version of the library, we used DirectSound to perform this function, which imposed certain restrictions on our looping behavior. All sounds looped from start to finish, regardless of the points specified in the loop data structure. In this custom mixer version of the Soundtrack Manager, we expand our SAMPLE looping capabilities, which requires a slightly expanded SAMPLELOOPINFO structure:

```
typedef struct _loop_t {
  BOOLEAN    active;
  UINT32     point1;
  UINT32     point2;
  BOOLEAN    pingPong;
} SAMPLELOOPINFO;
```

Streaming sounds make use of both point1 and point2. If they are both set to 0, the entire sound will be looped. You can adjust the end point of the loop by setting `point2` to the desired sample number. The `pingPong` parameter has no effect on streaming sounds.

The loop behavior for non-streaming sounds has also changed. If the `active` flag is set in `SAMPLELOOPINFO`, the sound will play all the way through and then loop back to `point1`. The first part of the `SAMPLE` between the beginning of the sound and `point1` is considered its attack phase, and doesn't get included in the loop. Furthermore, if `pingPong` is `TRUE`, the sound will play forward to the end, backward to `point1` and then forward again, over and over.

This whole topic of looping can be confusing, especially when the behaviors differ between Soundtrack Manager versions (as they do here). The bottom line is that we can loop either streaming or non-streaming sounds end-to-end without a problem in either version of the library we present in this book. We extend the looping behavior of both types of sounds with our custom mixer. Non-streaming sounds can now have a separate attack portion that will not be included in the loop. This is very important perceptually for those `SAMPLE`s used as MIDI instrument tones. We can also ping-pong the looped portion of non-streaming sounds. For streaming sounds, we now have the ability to specify the end point of the loop.

The final member of our new `SAMPLE` is a modified `SAMPLEMIDIINFO` structure:

```
typedef struct _samplemidiinfo {
  BOOLEAN isDrum;
  struct _sample_t *mutEx;
  SAMPLEENVELOPEINFO envelope;
} SAMPLEMIDIINFO;
```

As in the last chapter, we can cause the Soundtrack Manager to ignore all note off messages for this `SAMPLE` by setting the `isDrum` member to `TRUE` with the `SampleSetDrum` function. To simulate percussion sounds that are mutually exclusive, we use the `mutEx` field to identify another `SAMPLE` whose playback should be terminated when this one is played. And finally, we can directly manipulate the attack and release times of a `SAMPLE` via the `SAMPLEMIDIINFO` envelope structure:

```
typedef struct _envelope_t {
  UINT32 attackTime;
  UINT32 releaseTime;
} SAMPLEENVELOPEINFO;
```

Move over, SAMPLEs

In this version of the Soundtrack Manager, we focus more on `VOICE`s than on `SAMPLE`s. For instance, in the `PlaySample` routine we call `GetVoice` to find a free `VoiceTable` entry for this `SAMPLE`. In that routine, we start that free `VOICE` playing by setting the `isPlaying` member to `TRUE`. This flag is checked inside the mixing procedure as it loops through the `VoiceTable` and if set, processes that `VOICE`'s

SAMPLE data. The `PlaySample` function itself now returns a `VOICE` instead of a `SAMPLE`, and we specify that `VOICE` to the `StopSample` routine if we want to stop the sound before it's finished. These are the only two changes to our public API for this version of the Soundtrack Manager. The sample code to play a `WAV` file now looks like this:

```
SAMPLE *sample = NULL;
VOICE *voice = NULL;
if(sample = GetSampleFromFile("MyFile.wav", FALSE,
    FALSE, NULL)) {
  voice = PlaySample(sample);
  VoiceSetVolume(voice, 127, -127);
  VoiceSetFreq(voice, 23361);
} else {
  //Error loading WAV file
}
...
StopSample(voice);
```

You'll notice that immediately after calling `PlaySample`, we call `VoiceSetVolume` aned `VoiceSetFreq`. These are two new public functions in `PCMDev.c` that set the volume, pan and sampling frequency of the specified `VOICE`. If we're running the mixer in mono, the pan parameter has no effect and the volume of the `VOICE` is set to the specified value. Otherwise, the left and right volumes are calculated according to the specified pan position. We perform all sample-rate conversion for pitch-shifting and transposition inside our mixer as well. `VoiceSetFreq` sets the fixed-point `sampleIncrement` member of the specified `VOICE`.

Voice effects

`PCMDeviceSetDelay` and `PCMDeviceSetFlange` are two public functions defined in `\Code\Sound\Win9x\Win9xMix\FP\PCMDev.c`. These did nothing under DirectSound in Chapter 4, but here set some simple reverberation and flange parameters for the custom mixer to use. Reverberation is the result of the many reflections of a sound that occur in a room. We can now simulate reverberation by setting the volume, millisecond delay and feedback strength parameters to `PCMDeviceSetDelay`. Flanging is created by mixing a signal with a slightly delayed copy of itself, where the length of the delay is constantly changing. This is accomplished in the `PCMDeviceSetFlange` routine by specifying the volume, delay, frequency and depth parameters. The feedback signal can also be inverted by setting the `invert` flag to `TRUE`.

Mixing it up

Earlier in this chapter, we presented the four new mixer routines we wrote to relieve DirectSound of those mixing duties. All four of these functions are written in

floating-point assembly language for the PC, and are located in `\Code\Sound\`
`Win9x\Win9xMix\FP\Voicea.asm`. We select the specific mixing function to
use at the top of `PCMDev.c` (in this same `FP` directory) via three static variables: `CurrentOutputFreq`, `CurrentOutputBitWidth` and `CurrentOutput-`
`NChannels`. When the application calls `SMActivate`, this calls `DevicesActivate`
which in turn calls `PCMDeviceActivate` (if we've defined `__SLIB_DEVICE_PCM__`
in `modules.h`). After creating the DirectSound object and primary buffer, we call
`PCMDeviceSetOutputFormat` with these parameters to set the output format of
the new primary buffer:

```
if(PCMDeviceSetOutputFormat(CurrentOutputFreq,
  CurrentOutputBitWidth, CurrentOutputNChannels)) {
    Devices[SMInfo.PCMDevice].active = TRUE;
    return(PCMDEVICE_OK);
  } else {
    code = PCMDEVICE_UNSUPPORTEDFORMAT;
}
```

Inside this routine, we fill out a Windows multimedia system `PCMWAVEFORMAT` structure and call the primary buffer's `SetFormat` command. We reset our own internal
mixing parameters, and assign the correct `MixProc` to our `PCMdevice` function
pointer:

```
if(DS_OK ==
  Pdev->lpDSPrimaryBuffer->lpVtbl->SetFormat(
    Pdev->lpDSPrimaryBuffer, (LPWAVEFORMATEX)&PCMwf)) {
  //Reset mixing parameters
  Pdev->DSMixBuffer.lastUpdatePos = 0;
  Pdev->DSMixBuffer.nextWriteOffset = 0;
  //Choose appropriate mixing routine
  if(bitWidth == 8) {
    if(nChannels == 1) Pdev->MixProc = VoiceMix8Mono;
    else Pdev->MixProc = VoiceMix8Stereo;
  } else {
    if(nChannels == 1) Pdev->MixProc = VoiceMix16Mono;
    else Pdev->MixProc = VoiceMix16Stereo;
  }
```

Next we need to create the DirectSound secondary buffer into which our mixed
sounds will be placed. After setting another `PCMWAVEFORMAT` structure, we call
`CreateSoundBuffer` to do just that:

```
if(DS_OK ==
  Pdev->lpDirectSound->lpVtbl->CreateSoundBuffer(
    Pdev->lpDirectSound, &dsbdesc,
```

```
    &Pdev->DSMixBuffer.lpDSB, NULL)) {
} else {
  //Error creating DS mixing buffer
}
```

Call in the mediator

We could let our mixer routines mix directly into this DirectSound secondary buffer, but we've found this to be problematic. DirectSound requires us to lock and unlock this buffer when giving it more data to play. However, the updates provided by DirectSound as to how much data it has played since last time we checked are not always accurate. Furthermore, our underlying timer interrupts are sometimes delayed under Windows. For these two reasons, we have chosen to mix all our sounds into an intermediate buffer of our own creation, and then simply copy that data into the DirectSound secondary buffer for playback. This accommodates Windows' quirky behavior, and makes our mixing process more reliable and less prone to error.

We define __PCMDEVICE_MIXTOINTERMEDIATEBUFFER__ to signal that we wish to create and mix into an intermediate buffer. Within PCMDeviceSetOutputFormat, we create this additional buffer, and clear it out based on the mixer bitWidth. We also fill the DirectSound secondary buffer with silence to avoid any spurious noise when we tell the buffer to start playing:

```
#ifdef __PCMDEVICE_MIXTOINTERMEDIATEBUFFER__
  //Free any existing intermediate buffer
  if(Pdev->DSMixBuffer.intermediateBuffer.data) {
    SMFreeMem(Pdev->DSMixBuffer.intermediateBuffer.data);
  }
  intermediateBufferNBytes =
    Pdev->DSMixBuffer.bufferSegSize * 2;
  if(Pdev->DSMixBuffer.intermediateBuffer.data =
    SMGetMem(intermediateBufferNBytes)) {
      Pdev->DSMixBuffer.intermediateBuffer.size =
        intermediateBufferNBytes;
      Pdev->DSMixBuffer.intermediateBuffer.curFillPos = 0;
      Pdev->DSMixBuffer.intermediateBuffer.curPlayPos =
        intermediateBufferNBytes -
        Pdev->DSMixBuffer.bufferSegSize;
      if(bitWidth == 16) {
        APPMEMSET(
        Pdev->DSMixBuffer.intermediateBuffer.data, 0,
        intermediateBufferNBytes);
      } else {
        APPMEMSET(
        Pdev->DSMixBuffer.intermediateBuffer.data, 0x80,
```

```
        intermediateBufferNBytes);
    }
  if(Pdev->DSMixBuffer.lpDSB) {
    //Initialize mix buffer
    if(DS_OK ==
      Pdev->DSMixBuffer.lpDSB->lpVtbl->Lock(
        Pdev->DSMixBuffer.lpDSB, 0,
        Pdev->DSMixBuffer.bufferSize,
        &((LPVOID)lpWrite1), &dwLen1,
        &((LPVOID)lpWrite2), &dwLen2, 0)) {
      //Fill DS buffer with silence
      if(bitWidth == 16) {
        APPMEMSET(lpWrite1, 0, dwLen1);
      } else {
        APPMEMSET(lpWrite1, 0x80, dwLen1);
      }
      //Unlock DS memory
      Pdev->DSMixBuffer.lpDSB->lpVtbl->Unlock(
        Pdev->DSMixBuffer.lpDSB, (LPVOID)lpWrite1,
        dwLen1, (LPVOID)lpWrite2, dwLen2);
```

For the sake of completeness, the Soundtrack Manager still supports mixing directly into the DirectSound secondary buffer. In this case in our initialization, we simply lock the buffer, and mix in some initial silence:

```
#else
  if(Pdev->DSMixBuffer.lpDSB) {
    //Initialize mix buffer
    if(DS_OK ==
      Pdev->DSMixBuffer.lpDSB->lpVtbl->Lock(
        Pdev->DSMixBuffer.lpDSB, 0,
        Pdev->DSMixBuffer.bufferSize,
        &((LPVOID)lpWrite1), &dwLen1,
        &((LPVOID)lpWrite2), &dwLen2, 0)) {
          //Mix voices into this buffer
          Pdev->MixProc(lpWrite1, dwLen1);
          //Unlock DS memory
          Pdev->DSMixBuffer.lpDSB->lpVtbl->Unlock(
          Pdev->DSMixBuffer.lpDSB, (LPVOID)lpWrite1,
          dwLen1, (LPVOID)lpWrite2, dwLen2);
#endif
```

If all goes as planned, we start the DirectSound buffer playing and we're off:

```
//Start DS buffer
dwPlayFlags = DSBPLAY_LOOPING;
```

```
if(DS_OK == Pdev->DSMixBuffer.lpDSB->lpVtbl->Play(
  Pdev->DSMixBuffer.lpDSB, 0, 0, dwPlayFlags)) {
    PlatformExitCriticalSection();
    return(TRUE);
} else {
    //Error playing DS buffer
}
```

You will recall that to use this new mixer and play back SAMPLEs, all we have to do is call PlaySample. This calls GetVoice which sets the isPlaying member of the free VOICE structure to TRUE. Then on every timer interrupt, SoundManager calls UpdateDevices which in turn calls PCMDeviceUpdate. The pointer to the PCMdevice is extracted from the global Devices array which allows us to finally call MixProc to mix all playing VOICEs.

Summary

In this chapter we have written a custom mixer to replace the services of DirectSound. We have detailed the impact of this mixer on the Soundtrack Manager's internal structures and API. By doing this, we have increased our efficiency, and given ourselves more time for additional processing. Between the enhanced looping capabilities, the drum behaviors in the SAMPLEMIDIINFO structure, the attack and release fades specified in our SAMPLEENVELOPEINFO structure, the mixer's pitch-shifting abilities and the new reverberation and flange processing, we have created a powerful tool for the control and natural presentation of DA sounds in games.

9 Dynamic audio behavior – give me a call …

The individual components of the sound engine presented so far are brought together in this chapter. Dynamically controlling these subsystems is necessary for musical responsiveness. It is shown how procedure and play lists, callback mechanisms and periodic update routines help put into action such dynamic behavior.

Keep me movin'

Sound is a phenomenon experienced in time. It is always moving, from where it's been to where it's going next. There is no such thing as a naturally occurring, stationary aural "snapshot." Digital audio comes close to this by sampling a continuous waveform every so often, but this is not how we as humans experience sound. We perceive sound as pressure variations or waves through some observed point. Stop those pressure fluctuations, and there is silence.

The same can be said of sound production. All musicians, be they singers, string players, percussionists, keyboardists, woodwind or brass players must set something into motion to make their music. Conductors are always moving as well, guiding the live performance of a piece of music with his or her baton, hands, arms, facial expressions and at times their whole body.

Therefore, it is no surprise that the Soundtrack Manager is also an ever changing and moving temporal entity. It must pay attention to the passage of time to create and regulate a game's dynamic music and sound behavior. It functions as both the conductor and the orchestra, sometimes following orders from an external entity, at other times marching to the beat of its own drummer. While not literally alive, it must behave as such and be ready at any time to interpret and execute the sonic instructions it receives.

On your mark, get set . . .

A lot goes into the setup of the Soundtrack Manager to support this timely behavior. Much of it is akin to the preparations before a concert or play. Following this analogy, the performers start to arrive long before the audience is allowed into the auditorium. Depending on their role, people get instruments out of their cases, tune up, warm up their fingers or voices or lips, put on makeup and change into their costumes. The stage crew sets up the chairs and stands, plugs in the amplifiers, tests all the microphones and speakers, set their levels and label all the faders on the mixer. As it gets closer to curtain time, everyone gathers for those last few notes. Finally, the house is opened, everyone takes his or her places, and the show begins.

Within the Soundtrack Manager, all setup flows from the `SMInit` routine (in `\Code\Sound\SM.c`). The memory manager (`InitMemMgr`) and master volume pointers are initialized (`InitVolumes`). MIDI handler functions, the voice table, and all of our defined devices are set up (`InitDevices`). For each of the included modules, we arrange for such things as global `SAMPLE` and `MASTERPLAYER` lists, note tables, scripts (see Chapter 10), proximity sounds (in Chapter 11) and the CD player by calling its initialization function (via the function pointer `moduleInitProc`). We also build several lists of module-specific function pointers, called procedure lists, to activate, deactivate, update and uninitialize each module over the course of the Soundtrack Manager's operation. Once these lists are assembled, we don't have to remember what modules we're using in those situations. For each of the above operations, we simply traverse the correct procedure list, and all necessary action is taken.

Go!

It's final notes time, now. With the high-level infrastructure in place, we effectively raise the curtain on the Soundtrack Manager by calling `InitPlatform` to start a periodic timer. This is the thing that drives forward our entire audio operation from this moment. When the timer's period expires, it calls the main `SoundManager` function (also in `\Code\Sound\SM.c`) to play out all the sound in this frame:

```
void SoundManager(void)
{
  BOOLEAN ok=FALSE;
  ++InCriticalSection;
  SMInfo.frame++; //Increment frame counter
  do {
    ok = UpdateDevices();
    UpdateVolumes();
    ProcListExecute(UpdateProcList);
    UpdatePlatform();
  } while (ok);
  --InCriticalSection;
}
```

`SoundManager` calls `UpdateDevices` to read and process any MIDI input messages, and to play out all DA data both from the wavetable synthesizer (see Chapter 7) and other non-MIDI controlled SAMPLEs (see Chapter 4). We adjust all our master volumes (UpdateVolumes), and bring all selected modules up-to-date by executing the functions in the UpdateProcList. Finally, some platforms require an additional operation or task to be performed. We call this function in UpdatePlatform, and go to sleep until the next timer interrupt.

The SM_TIMER_PERIOD parameter (defined in \Code\Sound\SMPlatfm.h) sets the timer period in milliseconds. On the PC this is set to 8 ms, which translates into 125 updates per second. Anywhere in the 5–10 ms range is acceptable. Much faster, and the processing overhead of the timer starts to approach the timer period itself. Some platforms do not allow a period less than 1 ms. Above 10 ms, and people begin to hear quantization in the output sound, especially when playing MIDI files. There is a trade-off between processing time and musical accuracy to be considered when setting this parameter. If all you're doing is playing back straight DA, you can stay to the high side of this range. If you're using the full facilities of the MIDI wavetable synthesizer, however, a shorter value is better. There is no hard and fast rule for this, and the optimal setting will change across platforms.

So in summary, here's how the Soundtrack Manager works: the timer interrupt service routine periodically calls SoundManager. SoundManager updates the devices and calls all the functions in UpdateProcList. Each of those routines updates one of the included modules selected when we built the library. And it is those update methods that keep our DA data pipelines filled, our MIDI files playing and everything sounding at the right volume and tempo.

Points of procedure

Losing and gaining focus

We've talked about some of the procedure lists constructed during the initialization phase of the library. Another of these is DeactivateProcList, which is called when the games loses focus. At this point, the game calls SMDeactivate to temporarily relinquish control of all the audio resources we've grabbed for the game.

```
BOOLEAN SMDeactivate(void)
{
  if(active) {
    if(PlatformDeactivate()) {
      if(DevicesDeactivate()) {
        if(ProcListExecuteB(DeactivateProcList)) {
          active = FALSE;
          return(TRUE);
        }
      }
    }
  }
```

```
  } else {
    return(TRUE); //Already inactive
  }
  return(FALSE);
}
```

This routine turns off all sounding notes in `PlatformDeactivate` and closes all defined devices in `DevicesDeactivate`. The functions in `DeactivateProcList` are then called to handle what each module needs to do to gracefully stop all playing sounds and get out of the way for the next application.

The same thing happens in reverse when the game regains focus. At this point, the game calls `SMActivate`. All module-specific re-enervation happens via the routines in the `ActivateProcList`. All underlying devices are reclaimed and re-opened in `DevicesActivate`, and any platform-specific restart operations are taken care of last in `PlatformActivate`.

Hasta la vista, baby

The last of these procedure lists gets called when the game is about to exit. In this situation, the game calls `SMUninit` (located in, you guessed it, `\Code\Sound\SM.c`):

```
void SMUninit(void)
{
  SMDeactivate();
  ProcListExecute(ExitProcList);
  ProcListFree(ExitProcList);
  ProcListFree(UpdateProcList);
  ProcListFree(ActivateProcList);
  ProcListFree(DeactivateProcList);
  UninitPlatform();
  UninitDevices();
  UninitVolumes();
  UninitMemMgr();
}
```

`SMDeactivate` is called to release all the audio resources as above, followed by all the functions in `ExitProcList`. This frees up all the memory we have allocated within each module along the way. Again, we don't have to remember what modules those are. We simply call each routine in that list, and we're assured of releasing everything. At this point in our shutdown code, we back out the same way we came in during `SMInit`. We free all procedure list memory, shut down the platform timer, release all the devices and volume pointers, report any memory leaks (depending on the definition of `__SLIB_MEMORYLEAK_DEBUG__`) and thank everybody for coming. The doors to the auditorium are closed, the lights are turned off and the game is but a pleasant memory.

Driving Miss data

One of the coolest things about the Soundtrack Manager is its dynamic behavior. It grabs and releases the resources it needs as it needs them. It allocates procedure lists, SAMPLE lists and MASTERPLAYER lists, NoteTables and VOICE lists, fills out FADEINFO structures and SAMPLEENVELOPEINFOs and KEYMAPCHANNELs and all manner of other things in response to what it is told to play and how it is supposed to play it. For the most part, these operations are performed under the control of the platform timer interrupt. But there are some actions driven by the data itself.

In the midiInOpen call, DirectSound allows you to write your own callback function to process messages sent by the device driver. Therefore, when the MIDIindevice is defined, we specify the address of the MIDIInCallBack function to handle all incoming MIDI messages. These messages can arrive at any time, or never. When they do, we place them into the static MIDIBuffer (in \Code\Sound\Win9x\MIDIdev.c). This buffer gets processed every timer interrupt in the ServiceMIDIBuffer routine. We keep a separate set of NUM_MIDI_NOTES VOICEOWNERs in this source file, but otherwise parse and interpret the incoming MIDI stream exactly as for MIDI files (see Chapter 7). Each MIDI channel is mapped to either the MIDI output or PCM device in the MIDIMapsInit routine (called from InitDevices in \Code\Sound\Win9x\Devices.c), and we use this channel mapping to call the appropriate MIDIHANDLERS for each channel's device.

Hey, where'd everybody go?

Occasionally we want to be notified when a particular situation presents itself. This is the idea behind the NoVoicesCallback in the VOICEOWNER structure (defined in Devices.h). Remember that the VOICEOWNER structure itself can change from platform to platform while its name remains constant. For example, on some platforms we don't directly control what voices get turned on in response to MIDI note on messages (e.g. on the PlayStation2). In these cases, we can include a field in VOICE-OWNER to keep track of what system voices get turned on for each note in the synthesizer. Should the system steal any of those voices, we can track that occurrence and free any of our own resources associated with those voices when they're all gone.

Wait until I'm done

At any one time in a game, there will likely be several sounds playing simultaneously. Within our update architecture, each of these sounds, mostly SAMPLEs, are controlled individually either through the Soundtrack Manager API or via the MIDI wavetable synthesizer. However, it is useful at times to synchronize game events to sound behaviors. For instance, at the end of a level there is typically some "LevelWon" or "LevelLost" music that plays before moving on to the next stage. In this case, the game should wait until that music is done before displaying the next level's graphics. The Soundtrack Manager provides just this kind of service via the `DonePlayingCB` parameter of the `GetSampleFromFile` API call. In the `DSMIX` version of our library, this function pointer is copied into the `DonePlayingCB SAMPLE` member when the file is opened. When `UpdateSamples` determines this sound is done, it calls `StopSample` to stop the DirectSound secondary buffer and issues that callback.

Here I am

Back in Chapter 5, we talked about a mechanism where the game could be synchronized with the audio playback. By registering a procedure to be called for every meta-event marker message in a MIDI file, strategically placing these messages throughout the MIDI file can drive slide shows or animations or any number of things. The procedure, called a `MIDIFILEMARKERPROC`, is registered by calling `MIDIFileSetMarkerProc` and passing a pointer to the function to be called. It returns the previously set marker procedure, if there was one. For example:

```
MIDIFILEMARKERPROC *previousProc;
previousProc = MIDIFileSetMarkerProc(myMarkerProc);
```

This callback, here `myMarkerProc`, gets notified whenever a marker message is encountered in the file. In this way, the music lets you know where it is as it plays without any elaborate and often inaccurate timing schemes. The prototype for this callback functions is defined in `\Code\Sound\MIDIFile.h`:

```
typedef void
  (*MIDIFILEMARKERPROC)(MASTERPLAYER *masterPlayer,
    UINT32 markerNum, char *markerName);
```

As shown above, the function receives a pointer to the `MASTERPLAYER` playing the `MIDIFILE`, the marker number and a pointer to a character string containing the marker name (given to it by the audio artist when preparing a script command for this feature). The callback has full access to the Soundtrack Manager public MIDI API via the `MASTERPLAYER` parameter. Either the game or the Soundtrack Manager itself can set and use this communication channel. In fact, we put this to great use in Chapter 10 when we support English-language text scripts created by the audio artist.

Playing the field (of lists)

The last piece of the Soundtrack Manager MIDI puzzle is playlists. A `PLAYLIST` is a linked list of MIDI files to play one right after the other. These are useful for creating long background music tracks out of several MIDI files without the need for large DA files or much tending. They can also be used to score individual scenes. Any number of PLAYLISTs can be created and played simultaneously or sequentially, and are subject to the same master commands and controls as individual MIDI files or SAMPLEs.

Taking a closer look, the PLAYLIST structure is defined in \Code\Sound\ Playlist.h to be:

```
typedef struct _playlist {
   BOOLEAN          isLooped;
   BOOLEAN          is Paused;
   PLAYLISTENTRY    *entries;
   PLAYLISTENTRY    *entryNowPlaying;
   PLAYLISTENTRY    *entryToSegue;
   UINT16           pan;
   UINT16           volume;
   struct _fadeinfo *fade;
   struct _playlist *next;
} PLAYLIST;
```

The isLooped and isPaused data members indicate whether the PLAYLIST is looping or paused. Three PLAYLISTENTRY pointers follow next. These contain pointers to the MIDIFILEs and MASTERPLAYERs used in this PLAYLIST:

```
typedef struct _playlistentry {
   struct _MIDIfile       *MIDIFile;
   struct _masterplayer   *masterPlayer;
   BOOLEAN                removeWhenDone;
   struct _playlistentry  *next;
} PLAYLISTENTRY;
```

The PLAYLIST entries member is the MIDIFILE content list itself. entry-NowPlaying points to the current PLAYLISTENTRY, while entryToSegue points to the PLAYLISTENTRY to be played next. Normally, the MIDI files play out in the order specified in the entries list, and each file plays through to the end. However, as we'll see below, it is possible to change the order of a PLAYLIST after it's started, or to jump to an arbitrary PLAYLISTENTRY immediately. It is also possible to automatically remove a given PLAYLISTENTRY from a PLAYLIST by setting its removeWhenDone member to TRUE. This is useful in those situations where an entire PLAYLIST is looped, but we only want to hear a given MIDI file once.

The final few members of the PLAYLIST structure are pretty self-explanatory. The pan and volume members save the pan and volume settings for this PLAYLIST. As is evidenced by the fade member, we can fade a PLAYLIST the same way we fade an individual MIDI file or SAMPLE. Finally, we can also concatenate PLAYLISTs by using the next member of the structure.

Public listings

The Soundtrack Manager offers an extensive public programming interface for using and managing PLAYLISTs. For the most part, PLAYLISTs are considered to be just another Soundtrack Manager resource. This abstraction is borne out in the PLAYLIST API that contains many of the same functions as provided for individual MIDI files or SAMPLEs. To be sure, PLAYLISTs present additional functionality that doesn't apply to singular entities. In those cases, the API is extended to accommodate the new behavior. The bottom line is that the interface is complete, easy to use and applies the high-level behaviors of the Soundtrack Manager to this new and important game audio resource.

The GetPlayList function allocates, initializes and returns an empty PLAYLIST structure, while the FreePlayList function cycles through all the PLAYLIST entries and frees all MIDI file MASTERPLAYERs. It then frees the fade structure, if it exists, before finally freeing the specified PLAYLIST. You can append or remove a MIDI file to a specified PLAYLIST using the PlayListAppendMIDIFile and PlayListRemoveMIDIFile functions. Because it is possible to add the same MIDI file multiple times to a PLAYLIST, the latter function takes a Boolean argument to indicate whether you want to remove only the first occurrence of the MIDI file, or all of them.

There are two public functions to delete one or more entries from a PLAYLIST. PlayListRemoveEntry allows you to specify a specific entry number from a PLAYLIST. There are a couple interesting cases to be handled here. If the entry to be removed is currently playing, PlayListRemoveEntry will let the file finish playing but set its removeWhenDone flag to TRUE. This defers the entry's removal from the PLAYLIST until it's done playing (sensed in UpdateMasterPlayer) and the function returns successfully. If the entry being removed is the one to which we're to segue, the entryToSegue member is set to NULL. Finally, the desired PLAYLIST-ENTRY is removed from the PLAYLIST and deleted. PlayListClearEntries removes all PLAYLISTENTRYs from the specified PLAYLIST. This allows you to rebuild a PLAYLIST without throwing it away in favor of a new one. Any defined fade member is also left untouched. If any entry is playing, that MASTERPLAYER is stopped before it is freed.

PlayListGetVolume and PlayListSetVolume get and set the PLAYLIST volume and pan parameters, respectively. If either the input volume or pan is different from the current settings for the PLAYLIST and some entry in the PLAYLIST is playing, PlayListSetVolume adjusts all the currently sounding notes to the new volume and pan settings. It does this by calling the appropriate SetNoteVelocity MIDI handler for each channel's device. This real-time adjustment of a note's volume and pan is made possible only by the Soundtrack Manager's new NMD MIDI file format (see Chapter 6) and the fact that we keep a record of all currently sounding notes, their channels and associated devices (see Chapter 7). This cannot be done using SMFs.

It is also possible to loop and fade an entire PLAYLIST. The PLAYLIST isLooped data member is retrieved and set using the PlayListGetLoop and PlayListSetLoop functions. PlayListSetFade behaves in the same way as all of the Soundtrack Manager's other API fade functions. Using the target volume, pan and millisecond timing parameters, this function sets the `PLAYLIST fade` member to fade the entire `PLAYLIST` in or out relative to the `PLAYLIST`'s current volume and pan. The final `stopWhenDone` parameter indicates whether the `PLAYLIST` should stop playing or not when the fade is completed.

As pointed out above, `PLAYLIST`s are treated in a similar fashion as individual Soundtrack Manager resources. The next few functions continue to bear this out. You can play, stop, pause and resume an entire `PLAYLIST` using the `PlayPlayList`, `StopPlayList`, `PausePlayList` and `ResumePlayList` API calls. But `PLAYLIST`s also present some additional functionality that we see in the final three members of the API.

`JumpToNInPlayList` is a way to jump cut to or start a `PLAYLIST` from an arbitrary entry. It calls `PlayPlayList` with the desired entry number. `JumpToNext-InPlayList` will immediately jump to the next entry in the `PLAYLIST`, if it's playing, and handles a number of interesting playback situations. If `entryToSegue` is defined, that entry is ignored for the next one in the list. Should the list be set to loop and we're at the end of the list, `JumpToNextInPlayList` will start over at the top of the list. If it is not looping and the last entry is playing, the list is stopped. Finally, if the list is not currently playing, this function will start the list playing from the first `PLAYLISTENTRY`.

The last function in the `PLAYLIST` API is `SegueToNInPlayList`. This function allows you to assign the `entryToSegue` field of a given `PLAYLIST` to a particular entry number. The specified segue entry can be in the same or a different `PLAYLIST` altogether. Many interesting audio progressions can be set up in the Soundtrack Manager using this functionality.

In combination with `MIDIFILEMARKERPROC`s and `DonePlayingCB`s, and all the power of the Soundtrack Manager `MIDIFILE` API, `PLAYLIST`s round out the powerful game audio asset that is MIDI.

SM now, Redux

We've talked quite a bit about how the Soundtrack Manager works, detailing its public API and internal operation. However, we've not laid out in any clear and concise fashion how a game or other multimedia application would start up and shut down the sound library. Below we present the code and accompanying description for doing just that for a Windows console application:

```
BOOLEAN SMDisplayErr(char *string)
{
  printf(string); return(TRUE);
}
HWND GetHWND()
{
```

```
        //Find console window and return its window handle
        return FindWindow("ConsoleWindowClass", GetTitle());
    }
    int main(int argc, char *argv[])
    {
      SMSetSYSStringHandler(SMDisplayErr);
      if(SMInit()) {
        SMSetHWnd(GetHWND()); //Get app window handle
        SMActivate();
        //Place game loop here.
        //Multiple calls to SMDeactivate and SMActivate as
        // game loses and regains focus.
        SMDeactivate();
        SMUninit();
      } else {
        //Failed to initialize sound engine
      }
      return(0);
    }
```

The first thing the application does is define a function to receive any text error messages from the Soundtrack Manager. This kind of error reporting is essential during application development, and can be turned off by defining SM_NO_ERROR_MESSAGES in SMPlatfm.h and recompiling just before you ship. In the example code above, SMDisplayErr will simply print out these messages.

The next function, GetHWND, is a simple routine to get the window handle for this sample console application. This is absolutely necessary on the PC if we're using any DA. This is because the DirectX Audio SetCooperativeLevel function, called from within PCMDeviceActivate during the SMActivate call, requires a pointer to the main window handle of the application. We pass this handle to the SMSetHWnd call just before activating our library.

Getting into the main routine, we tell the Soundtrack Manager we want to use our SMDisplayErr routine for all error reporting via the SMSetSYSStringHandler call. Next we initialize the Soundtrack Manager with a call to SMInit. This function has been described in great detail both here and in other places in this book (see Chapter 3). Based on the modules and devices defined in the modules.h header file, this routine sets up all the necessary memory, system and platform resources.

The SMSetHWnd function follows next to inform the Soundtrack Manager of the application's window handle which we discussed above. This is followed by a call to SMActivate that kicks the library into action. All of the defined devices are started within this routine, and the Soundtrack Manager is fully open for business.

The game is now running, and can issue any of the public Soundtrack Manager API calls to load and play back whatever audio is desired and necessary. When it loses and gains focus, it need only call SMDeactivate and SMActivate to gracefully switch the Soundtrack Manager off and back on.

When the game is about to end, the game issues one final call to `SMDeactivate` to stop and release all claimed audio devices. It then calls `SMUninit` to shutdown the library completely. This function calls all the functions in the `ExitProcList` to free each module's memory. It stops and frees the platform timer, frees any residual memory for the devices and volume procedure lists and finally shuts down the memory manager.

That's all folks!

Summary

In this chapter, we brought together the individual components of the sound engine. It was shown how these subsystems are dynamically controlled for musical responsiveness. It was also shown how procedure and play lists, callback mechanisms and periodic update routines help put into action such dynamic behavior. Finally, a quick overview of the API calls necessary to start, run and stop the Soundtrack Manager was given.

10 Giving control to the audio artist – I got the power!

We have seen how to use the Soundtrack Manager through its programming interface. But this is of limited use to an audio artist who is typically not a programmer. This chapter fulfills our last major objective: that the full power of the sound engine be available to an audio artist to create the game's soundtrack. A cue-based text interface providing direct control of all the audio engine's services is presented.

Handing it over

Way back in Chapter 3, our stated goal was to design and build an interactive audio system that allows the audio artist to construct a compelling soundtrack that will adapt and change in response to the player's actions in real-time and in an appropriate manner for the current game situation. The audio system we've presented in the preceding chapters goes a long way towards making the latter part of that goal a reality. Though it's taken us awhile to put the underlying pieces in place, we're ready to turn the Soundtrack Manager over to the composers and artists for which it is intended.

We accomplish this through a collection of text commands that mimic the Soundtrack Manager's high-level API. The composer assembles these commands into individual audio cues, defining detailed audio actions for specific game events. The file of cues, called a script, gets compiled into a custom binary format that is loaded into the game for the given scene or level. The audio artist and game programmer agree on what cue number to call for what game event, and the rest is up to the artist. The game calls the individual cues, but the audio action taken inside those cues is entirely up to the composer. Audio changes no longer require recompiling to be heard, so long as the cue numbers don't change. A new script with new files, and we're good to go.

There are several pieces to this audio scripting scenario. First there are the commands themselves that give the composer complete control of the Soundtrack Manager. Next comes a utility, called SNDEVENT, which converts the text script into a binary format for easier reading, parsing and searching by the engine. A simple interface between the game and the sound engine is needed to load the binary script file and call the cues. Finally, the Soundtrack Manager must read the binary script file and translate the desired commands into the appropriate function calls to execute the desired commands. In the following sections, we'll take a look at each of these pieces in succession.

The script commands

The first command is the C++-style comment delimiter, //. This allows you to insert comments into the script file, explaining your actions and labeling the cues. Everything after the delimiter on the same line is ignored when parsing the script. As a general rule, you can never have too many comments. These make the script easier to read, and are discarded in the binary compilation phase, add nothing to the final binary script file size. So go ahead and use them. For example:

```
// Beginning of script for Level 1
// Introduction and setup
```

The PRINT command is used to indicate the progress of the script compiler. Whatever text is enclosed in double-quotes after the PRINT command is directed to the console window of the SNDEVENT compiler application as it processes the script. Like comments, this text is also discarded in the final output file:

```
PRINT "Compiling intro cues"
```

The CUE command defines the beginning of a collection of Soundtrack Manager commands. Each CUE is identified by a number that is called by the game to execute some specified group of audio actions. All commands on following lines after the CUE keyword and number are executed when the cue number is called. The ENDCUE command terminates the CUE:

```
CUE 1
  PRINT "Compiling CUE 1"
  //Setup scene
ENDCUE
```

If the script is small, using numbers for CUEs is usually not a problem. However, as a script becomes larger and more complex, it can be difficult to keep all the numbers straight. This is where the DEFINE command comes in. It allows you to define meaningful text tokens for numbers. These tokens can then be used throughout the rest of the script instead of the numbers they represent. DEFINE statements make the script more readable and understandable, and can appear anywhere throughout the script:

```
DEFINE INTRO_0 1
define intro_1 11
```

```
define intro_2 21
CUE intro_0
  //Setup scene
ENDCUE
```

The RESET command unloads all the resources and structures loaded by any previous scripts. Any MIDI files, PLAYLISTs or SAMPLEs are stopped, unloaded and
freed. This command provides the audio artist with an easy way to wipe the audio
slate clean and start fresh:

```
DEFINE RESET_CMD 0
cue reset_cmd
  //Unload all script audio resources
  RESET
endcue
```

The next set of commands have to do with loading and playing DA. In the context of
a script, a piece of DA is considered a sound effect. The first command is LOADSFX.
This command takes a maximum of six arguments, though only the first two are
required. The arguments are: the sound effect number, the relative path name to
the DA data file, the sound's volume and pan, its master music setting and a flag
to stream the data or not. For instance, the following LOADSFX command loads
the growl.wav file from the sfx directory, and calls it sound effect 0. It is set to
maximum volume (127), panned center (0), is not part of the music soundtrack (0),
and should be streamed from disk (1):

```
CUE 10
  //Load monster growl
  LOADSFX 0 sfx\growl.wav 127 0 0 1
endcue
```

All sound resources in scripts begin counting from 0. The volume parameter can take
on any value from 0 (silence) to 127 (maximum gain), while the pan parameter
ranges from −128 (far left) to 0 (center) to 127 (far right).

Regarding file names in commands, the Soundtrack Manager keeps a global variable called SoundPath. This can be set to any directory by the game, and specifies
where the sound resources are stored in the file system. Commands such as
LOADSFX specify the relative path name to the specific resource starting from
SoundPath. In this way, the audio artist can group their resources in any fashion they
choose, making and populating any number of directories from SoundPath on
down. The files and folders are handed off to the game programmer, and referenced
in the script by their relative path names. While this makes life easier for the audio
artist by cutting down on the typing necessary within the script itself, the implications
of this method of audio content delivery and organization are much more significant.
The audio artist can reconfigure and change the content of the various files and

folders completely independent of the programmer. The person or persons who create the audio content retain full control of its substance, organization and use. When new sounds are needed or ready, the audio artist simply supplies a new batch of audio files and scripts. The game calls the CUEs as before, and the audio changes just happen.

Many different sounds can be used to construct a game's background music track. It would be nice to treat all those sounds as a whole instead of having to keep track of them individually. This is the idea behind the master music parameter in this LOADSFX command above (and others as we'll see later on). Within the Soundtrack Manager, we consider music separate from sound effects and have a separate volume and pan control for music (see the MusicMasterVolumePtr structure in \Code\Sound\Volume.c). Sounds can be flagged to be part of the music track when they're loaded. Thereafter, the volume and pan of all those sounds can be adjusted simultaneously using the MUSICVOLUME and FADEMUSIC script commands (described below). Similarly, the volume and pan of those sounds not designated as being part of the music track can be manipulated en masse by using the SFXVOLUME and FADESFX script commands (which use the SFXMasterVolumePtr in \Code\Sound\Volume.c).

Many script commands contain optional arguments that get set to default values if they're not present. In the case of LOADSFX above, only the desired sound effect number and file name are required. If the remaining parameters are not supplied, the sound's volume is set to maximum, panned center, is not part of the music track and is fully loaded into RAM for playback. However, you cannot skip some optional arguments and specify others. The parser is not smart enough to know what arguments have been skipped and which ones have not. Therefore, you must supply all intervening optional arguments up to and including the one you wish to specify.

A few additional comments on script formatting. As suggested in the examples above, scripts are case-insensitive, and must reside on a single line. This includes comments, which must also be on their own line. Each command argument must be separated by at least a single white space character. Tabs are also allowed, but other delimiters such as commas or semicolons are not. Finally, continuations across lines are not permitted.

Complementing LOADSFX is UNLOADSFX. This command takes a single parameter, the sound effect number to unload. In one simple command, it removes the sound from the Soundtrack Manager's global linked list of SAMPLEs (see Chapter 4) and frees all its associated memory:

```
CUE UNLOAD_MONSTER
  //Unload monster growl
  UNLOADSFX 0
ENDCUE
```

The next four commands all take a single required parameter, the sound effect number: PLAYSFX, STOPSFX, LOOPSFX and UNLOOPSFX:

```
CUE 100
  LOADSFX 2 sfx\BFG.wav 127 0 0 0
```

```
    LOOPSFX 2
ENDCUE
CUE 101
   PLAYSFX 2
ENDCUE
CUE 102
   STOPSFX 2
ENDCUE
CUE 103
   UNLOOPSFX 2
ENDCUE
CUE 104
   UNLOADSFX 2
ENDCUE
```

The FADESFX command takes four parameters: a sound effect number, a target volume and pan location, and the length of the fade in seconds. Only the sound effect number is required. If not specified, the target volume is set to 0 (silence), as is the target pan (center). The length of the fade can range from a minimum of 1 s to a maximum of 255 s, and defaults to 4 s if not supplied:

```
define LOADANDFADEINBG1 200
define BG_MUSIC1 10
CUE LOADANDFADEINBG1
   LOADSFX BG_MUSIC1 music\BG1.wav 0 0 1 1
   //Fade in background 1 to max volume over 10 seconds
   FADESFX BG_MUSIC1 127 0 10
ENDCUE
```

The SFXVOLUME command takes three parameters: the sound effect number, volume and pan location. Again, only the sound effect number is required. If not specified, the volume is set to 127 (maximum) and the pan to 0:

```
CUE 201
   //Set BG1 music volume to half, half-left
   SFXVOLUME BG_MUSIC1 64 -64
ENDCUE
```

All Soundtrack Manager volume commands will set the volume and pan of the resource whether it's currently playing or not. If the sound is playing, the volume is immediately adjusted. Otherwise, it will use save those volume and pan settings for when it next starts up.

The next group of script commands have to do with MIDI files. LOADMIDI takes five arguments: the MIDI file number to which it is to be assigned, the relative path name of the NMD file to load, its desired volume and pan and a flag indicating whether the file should be looped or not. As with the `LOADSFX` command, only the number and file name are required. The volume defaults to 127, the pan to 0 and the looping flag to F (false):

```
CUE 120
  //Load level MIDI files - all looping
  LOADMIDI 1 music\Sept1.nmd 100 0 T
  LOADMIDI 2 music\Battle.nmd 100 0 T
ENDCUE
```

The MIDI file numbers are separate and distinct from the sound effect numbers, and the files themselves are in the new NMD format (see Chapter 6). Notice again the relative path names of the files.

`UNLOADMIDI` functions very much like `UNLOADSFX`, and takes a single MIDI file number parameter. It stops the file if it is playing, unloads the MIDI file and frees the underlying `MASTERPLAYER`:

```
CUE 121
  UNLOADMIDI 1
ENDCUE
```

The next eight commands also require a single MIDI file number. Each command is self-explanatory: `PLAYMIDI`, `STOPMIDI`, `LOOPMIDI`, `UNLOOPMIDI`, `PAUSEMIDI`, `RESUMEMIDI`, `MUTEMIDI` and `UNMUTEMIDI`:

```
CUE 300
  PLAYMIDI 1
ENDCUE
CUE 301
  STOPMIDI 1
  PLAYMIDI 2
  LOOPMIDI 2
ENDCUE
CUE 302
  UNLOOPMIDI 2
  PLAYMIDI 3
Endcue
Cue 303
  Pausemidi 3
  Mutemidi 2
Endcue
Cue 304
  Resumemidi 3
  Unmutemidi 2
Endcue
```

The `MIDIVOLUME` command takes a MIDI file number, volume and pan setting. If the volume parameter is omitted, the MIDI file is set to maximum volume. Likewise if the pan parameter is absent, the file defaults to the center position upon playback. Only the MIDI file number is required:

```
DEFINE START_SWORD 310
DEFINE PLOT_SWORD 311
DEFINE DLG_SWORD 10
DEFINE MIDI_SWORD 30
CUE START_SWORD
  LOADSFX DLG_SWORD music\Dialog_Sword.wav 123
  LOADMIDI MIDI_SWORD music\Swordplay.nmd 100 97 T
  PLAYMIDI MIDI_SWORD
ENDCUE
CUE PLOT_SWORD
  //Duck music for dialog
  MIDIVOLUME MIDI_SWORD 32
  PLAYSFX DLG_SWORD
ENDCUE
```

Just before `CUE START_SWORD` we define a number of tokens to make the script more readable. The cue loads the dialog chunk at a volume of 123, defaults to center pan, is not considered a music event and will not be looped or streamed. The background music is loaded at volume 100, panned quite far right and will loop as it plays. Cue `PLOT_SWORD` ducks the volume of the background track and plays the previously loaded WAV dialog file.

The `FADEMIDI` command takes a MIDI file number as its first and only required parameter. The next four optional parameters are the same as for the `FADESFX` command above: volume, pan and duration in seconds. These default to 0, 0 and 4, respectively:

```
CUE StartAndFadeLevel9
  LOADMIDI Level9Out music\CoolFade.nmd 127
  PLAYMIDI Level9Out
  FADEMIDI Level9Out 0 0 60
ENDCUE
```

This command loads and starts a MIDI file playing. It then fades the music out slowly over the next 60 s.

Many of the commands for MIDI files have track-based counterparts. They must each specify the MIDI file and track number on which they're operating, followed by various optional parameters. `MIDITRACKVOLUME` sets the volume and pan of the given MIDI file track. As with all other fade commands, `FADEMIDITRACK` sets the target volume, pan and duration of the specified MIDI track and proceeds to ramp that track up or down in volume within the time allotted. `MUTEMIDITRACK` has no optional arguments, and immediately mutes the specified MIDI file track. UNMUTEMIDITRACK takes an optional volume and pan parameter that allows you to turn the track back on in a different position than where it was muted.

In Chapters 5 and 6, we talked quite a bit about MIDI file markers. Formally, these are Standard MIDI File Marker Meta-event messages that can be inserted into a MIDI file using just about any MIDI file editor. The audio artist can insert any number of these meta-event marker messages into the MIDI file at musically significant locations, and can specify any desired text name for each event. During playback, the Soundtrack Manager notifies the game or itself of these markers by setting a MIDI-FILEMARKERPROC callback routine using the MIDIFileSetMarkerProc function (in \Code\Sound\MIDIFile.c). The code sends back both the sequential number of the marker as well as the text name given to it by the audio artist above. As seen in Chapter 6, we can also seek to these markers. This is a very powerful way of moving through and jumping around in MIDI files, even while they're playing, and is only possible because of the new NMD format.

SEEKTOMIDIMARKER is the script command to seek to a given marker within a MIDI file. It takes a MIDI file number, and then either a marker number or name. Most artists use the marker's name so they don't have to keep track of the individual marker numbers, especially if there are a lot of them. When the Soundtrack Manager receives this command, it finds the marker number corresponding to the given marker name and seeks through the file to that designated marker.

SETMIDIMARKEREVENT is used to select a script command to execute when the specified MIDI file next hits a marker event. It has two required arguments, the MIDI file number and the script cue number to call. This is an extremely useful interactive mechanism because it allows you to change the behavior of the audio on the fly, and do so on musically significant boundaries. For instance, you could be looping within a battle scene, but when the player finally kills the monster, the game can call the cue number containing the "monster killed" music at the next marker of the battle music. The audio artist can change the audio to whatever is necessary to end the battle, and be assured of doing so musically because of the marker events he or she inserted into the MIDI file upon its creation. This behavior is made possible by the fact that InitScript sets an internal marker procedure, ScriptMIDIFileMarkerProc, to be called whenever a MIDI file hits a marker (in \Code\Sound\Script.c). From a MIDI handling point of view, ScriptMIDIFileMarkerProc gets called by HandleMIDIFileEventMarker (in \Code\Sound\MIDIFile.c) for each marker event seen in the NMD file.

In the LOADSFX command above, we saw how to indicate whether the given piece of DA was a part of the game's music or not. The MUSICVOLUME command lets you set the volume and pan of all MIDI files and any flagged sound effects files simultaneously. If the sounds are currently playing, their volume is immediately updated. If they're not playing, their volume and pan settings are adjusted when they start. The command has no required parameters, and will default to a maximum volume (127) and center pan (0) if not explicitly included.

The FADEMUSIC command works the same way as all of our previous fade commands. There are no required parameters, but the artist can optionally specify the target volume, pan and duration of the fade. If not specified, the music will fade to silence in the center over 4 s. All MIDI files and tagged sound effects will be affected by the FADEMUSIC command. This is a very powerful and convenient way to control all the music at once without having to remember what is currently playing or not.

The next group of script commands has to do with MIDI file playlists. The first command, APPEND, takes two required parameters: the loaded MIDI file number and the desired playlist number to which to it should be appended. If the playlist does not currently exist, it is created. Then the specified MIDI file is appended onto that list. Any number of MIDI files can be added to any number of playlists, so long as they have been previously loaded and numbered using the LOADMIDI script command. There are practical limits imposed here, however. As of this writing, only 128 MIDI files can be loaded at any one time, and the current limit on the total number of playlists is 128 as well. These hard limits can be changed by a programmer (in \Code\Sound\Script.c) and the library rebuilt if these limits are too constraining.

The REMOVE command also takes the same two required parameters: the loaded MIDI file number and the desired playlist number. It removes the specified MIDI file number from the given playlist.

The next seven commands all take a single required parameter, the playlist number: PLAYLIST, STOPLIST, PAUSELIST, RESUMELIST, LOOPLIST, UNLOOPLIST and CLEARLIST. Each command behaves as you would expect to play, stop, pause, resume, loop, unloop or clear the given playlist.

We can also set the volume of a playlist using the LISTVOLUME command. This has one required parameter, the playlist number, and two optional arguments: the desired volume and pan position. All notes of a playing MIDI file list entry are immediately adjusted to reflect the new volume using the SetNoteVelocity MIDI handler for the appropriate device. Again, we can do this because of the new NMD MIDI file format and that we keep a record of all notes currently playing on each track for each MIDI file. The volume defaults to maximum and pan to center.

As you have probably already guessed, there is also a FADELIST command to fade the specified playlist to the desired volume and pan over a given duration. Here again, the only required argument is the playlist number. The three other optional arguments of volume, pan and duration default to 0, 0 and 4, respectively. The behavior of the fade is extended here to cross MIDI file boundaries as the playlist plays. As one file stops and another starts, the current playlist volume and pan settings are continuously applied.

The final two playlist commands allow you to move around the playlists in nonsequential fashion. SEGUETO takes three required arguments: the MIDI file number to play, the playlist number to which to segue and the currently playing playlist number to segue from. The source and destination playlists can be the same. Under normal circumstances, a playlist plays the MIDI files in succession. SEGUETO allows you to dynamically change the order of the list's playback both within and among playlists by specifying what MIDI file number to play next when the current MIDI file is done. For instance, the first cue below sets up two playlists and starts the first one playing and looping. The second cue causes the first playlist to segue to the third MIDI file in the other list, for whatever entry it may be playing at the time:

```
define StartLists 400
define SetNextListSegue 401
define destList 1
define srcList 2
CUE StartLists
  Loadmidi 1 music/BG1.wav
  Loadmidi 2 music/BG2.wav
  Loadmidi 3 music/BG3.wav
  Loadmidi 4 music/BG4.wav
  Loadmidi 5 music/BG5.wav
  Append 1 srcList
  Append 2 srclist
  Append 3 destlist
  Append 4 destlist
  Append 5 destList
  PLAYLIST srcList
  LOOPLIST srcList
ENDCUE
CUE SetNextListSegue
  SEGUETO 3 destList srcList
ENDCUE
```

The last playlist command is JUMPTO. This command lets you start playing a list at an arbitrary point or make a jump cut within a single playlist. It takes two required parameters, the MIDI file number and playlist number. Upon issuing this command, the specified MIDI file in the given playlist will start playing. If any other entry is playing in that list, it is stopped and the desired entry played. If the list is not currently playing, the list is started with the given MIDI file number.

The next batch of script commands allow the audio artist to access and use the facilities of the MIDI wavetable synthesizer presented in Chapter 7.

The MAPSFX command maps a previously loaded DA sound onto a range of keys on a specific MIDI channel. The command takes eight arguments, three required and five optional. The three required parameters are the loaded sound effect number, and the MIDI channel and the low or starting note number on that channel to which the sound will be mapped. The five optional arguments are listed next, as well as their default values in parentheses: the high note (value of the required low note parameter), center note (value of the previous high note), volume (127), pan (0) and a flag indicating the monaural status of the sound (false). Breaking these parameters out one-by-one, the sound effect number is the number of a previously loaded chunk of DA (see the LOADSFX command, above). The MIDI channel onto which the sound will be mapped follows next and will take on a value between 0 and 15. This is because in a standard MIDI configuration, there are 16 available channels, beginning with channel 0. The next three parameters define the low, center and high notes of the key range to which we're mapping this sound. The low and high notes specify the boundaries of the legal note range for this mapping. The sound will not respond to note on messages outside this range. The center note specifies the MIDI note number for which the sound experiences no transposition. For any note in the specified key range, an appropriate amount of transposition will be applied to the sound equal to the semitone difference between the note on message and the given center note. (Any channel pitch bend will also be applied to the notes). The volume and pan parameters are the same as previous script commands. The mono flag indicates that the sound is monaural, and is used only in the DSMIX version of the Soundtrack Manager. Stereo sounds can be loaded, but monaural sounds are typically used for the wavetable synthesizer.

For programmers, there is one code change of which to be aware when using the wavetable synthesizer. At the end of InitDevices, there is a call to MIDIMapsInit (in \Code\Sound\Win9x\Devices.c). This function calls InitializeMIDIMap that sets the desired device for each MIDI channel. Any channel that uses the wavetable synthesizer must set its MIDIMap entry to SMInfo.PCMDevice. The alternative setting is SMInfo.MIDIDevice, which uses the default MIDI out device of the platform. It is perfectly legal to mix and match what MIDI channels get assigned to which devices. When first starting out, it may make sense to use the native platform MIDI device to be sure you're seeing the MIDI stream you expect. This also makes sense if you don't want to go through the trouble of preparing custom DA sound banks before hearing any sound. There is also nothing wrong with using the MIDI out device if this suits your game's needs. Recall, however, that one of the main goals of the Soundtrack Manager is to make the game sound the same on all platforms. If that design feature is important to your game, you should plan on using custom sounds and mappings.

Layering is when more than one sound is mapped to the same MIDI channel and key. This is a very powerful technique as multiple sounds can be triggered with the same MIDI note on message, greatly increasing the perceived richness and timbral quality of the composite sound. The Soundtrack Manager accommodates layering, allowing up to MAX_SAMPLES_PER_KEY sounds for each key (defined in \Code\ Sound\Win9x\Keymap.h to be 2). Using the MAPSFX command, therefore, you can map up to two sounds to the same MIDI channel and key combination. Each of those sounds can also have their own envelopes and drum features (see below), as well as their own volume and pan settings.

The UNMAPSFX command takes two arguments, the sound effect number and the MIDI channel to which it was previously mapped. This command removes the specified sound from all notes on the given channel.

In Chapter 7, we spoke about how drums and other percussion instruments do not exhibit a well-defined cutoff and are generally left to resonate and fade away of their own accord. The DRUMSFX command tells the wavetable synthesizer to do just that by ignoring all MIDI note off commands for a given sound. It takes a single parameter, a loaded sound effect number. This assures us that the full drum sound will play and not be cut off prematurely by a MIDI note off message. UNDRUMSFX takes the same required parameter, a sound effect number, and turns this behavior off. Afterwards, all note off messages stop the sound whenever they occur.

Also in Chapter 7 we saw how the Soundtrack Manager wavetable synthesizer accommodates perceptually significant volume characteristics by directly manipulating the attack and release times of a sound. This capability is exposed in the script interface via the SFXENVELOPE command. This takes a total of three arguments: the sound effect number, and the millisecond attack and release times (both optional). The latter two parameters cause the sound to fade in and out over the specified durations in response to MIDI note on and note off command, respectively. They both default to 0 if not specified.

The PLAYMIDISFX command combines the Soundtrack Manager's MIDI interface with the wavetable synthesizer's sample mapping capability. It allows you to play individual notes of a key-mapped sound effect without using a MIDI file. It takes four arguments: two required, the MIDI channel and note number to play, and two optional, the desired volume and pan position. As always, the volume defaults to maximum (127) and the pan to center (0). Different notes result in different transpositions, limited only by the defined MIDI key ranges (see the MAPSFX command, above) and the transposition range of the Soundtrack Manager ($+/-3$ octaves). This is a very useful way to get extra mileage out of DA background tracks as you can start any number of instances of the sound playing at all different pitches and speeds.

The STOPMIDISFX command takes two required parameters, the MIDI channel and note number, and stops the sound mapped to that specific channel and key combination.

There are times in a game where circumstances change and audio cues that played earlier should not play now. This could be as straightforward as the loading of a level, which should only be done once, or when a player first crosses a threshold that starts some music playing, for instance. Should they cross that threshold again, the music should not restart. Another example could be at the conclusion of a chase scene, where the music for the chase should be stopped and disabled until the next time circumstances warrant. The Soundtrack Manager provides two cues to enable and disable any loaded audio cue (we'll talk about what that means a little later). This is another key instance where the responsibility of the audio track is kept in the hands of the audio artist. The game does not have to be aware of the state of the audio playback, and happily issues any sound events it has whenever and wherever they're placed. The format of the ENABLECUE and DISABLECUE commands is the same. They each take a single required argument, the cue number to allow or not, respectively.

There are also times in a game when a sound event should not start immediately, but play some time in the future. This is the idea behind the DELAYCUE command. It takes two required arguments, the cue number to delay and the delay time in milliseconds. It calculates the number of frames until the desired time of the event, and places a DELAYEVENT in a static double linked list called DelayEvents. On every timer interrupt, SoundManager calls the functions in UpdateProcList, one of which is UpdateScript. This routine calls UpdateDelayEvents that traverses the DelayEvents list looking for those cues it needs to play in the current frame. When it finds one, it fires it off, and deactivates the event for later housekeeping when the next sound cue is requested. The DELAYCUE command takes three additional optional parameters that are used to specify how far away the sound is (distance), the left/right pan position of the sound (balance) and the front/back position of the sound (fader). This is useful as you can make a sound appear or re-appear over a wide range of apparent locations. The distance parameter affects the volume of the sound, and takes an integer between 0 (close) and 255 (far away). The Soundtrack Manager plays the sound at full volume when distance is 0 (taking into account any additional master, sound effect or MIDI volumes). The sound gets progressively softer as the distance value increases, until it is all but silent at a distance of 255. The pan parameter specifies the sound's left/right position, and ranges from −128 (full left) to 127 (full right). The fader parameter specifies where the sound is located relative to the listener, either behind (−128), at the listener's head (0) or in front (127). Using a combination of these parameters, it is possible to make sounds move around pretty well in the horizontal plane without special three-dimensional sound processing. These three parameters are also in the SMEvent argument list, which is the main public API script function (see below). In the next chapter we'll make good use of these positional parameters when we talk about attaching sounds to game objects and dynamically coordinating their audio and visual display.

Currently, there is minimal support for using the native CD audio player on a given platform. The PLAYCDAUDIO command takes one required argument, the track number, and plays that track. The STOPCDAUDIO command simply stops that playback.

Within our script interface, there are several useful commands to handle major changes in the state of a game. The first one is PAUSEGAME. When the artist issues this command, the Soundtrack Manager pauses all currently playing MIDI files and builds a list of what files to restart later when we receive the RESUMEGAME command. This command spurs the Soundtrack Manager to traverse that saved list of MIDI files, and resume their playback exactly where they left off. All catchup events are processed, so all notes playing at that moment, including those notes started before the pause event was issued but which have not yet received a note off command, will sound.

The SAVEGAME command attempts to save a record of what files are playing at the moment the command is issued. The Soundtrack Manager has a bit more work to do here as it actually builds a special cue on the fly to restore the state of the currently playing MIDI files (called RELOADGAMECUENUMBER). RELOADGAME simply calls the RELOADGAMECUENUMBER cue, and all the MIDI files will restart.

The last two commands in the Soundtrack Manager's scripting interface give the audio artist access to the delay and flange effects processing in the FPMIX version of the library. The SETDELAY command sets the parameters of a global reverberation effect. It is beyond the scope of this book to define or describe how reverberation works. There are countless tomes on the subject, and the reader is encouraged to seek out more information if they so desire. The algorithm used in the Soundtrack Manager simply mixes in a delayed copy of the original sound and includes a single feedback path. The SETDELAY command takes three parameters, all optional. It allows you to set the volume of the delayed sound between 0 (silence) and 127 (full gain), the length of the delay (in milliseconds from 0 to 65535) and the feedback strength of a recirculated copy of the input sound (from 0 to 127). All three parameters default to 0 if not specified.

The SETFLANGE command takes five arguments, again all optional and defaulting to 0 if not specified. These include the volume of the effect (0 to 127), the length of the delay (in samples from 0 to 1000), the number of samples for the depth of the effect (again 0 to 1000), a nonnegative sweep frequency and a flag to indicate whether the recirculating sound should be inverted or not (T or F, respectively). Again, it is beyond the scope of this book to explain what constitutes a flange effect. Those who are interested in using this commands are encouraged to find out more about the effect as it can be quite tricky to use. Otherwise, play around with it, and hear what happens.

That wraps up the discussion of the script commands themselves. For the sake of clarity and ease of use, all the commands and their parameters have been collected in Appendix A.

It's all in the translation

Once a script file is created for the game, it needs to be compiled into a custom binary format for quick access during runtime. Included on the CD is a Visual C++ project for a Windows console program, called SNDEVENT, which performs this conversion. Three additional projects are included for the support libraries used by this application (Xbase, Xlparse and Xlnparse).

SNDEVENT converts the text script file into a binary script file by writing the information for each command into a RAM image of the output file on the fly. The bulk of the conversion work in this application is done in the MakeBinaryFromScript routine (in \Code\Apps\Sndevent\Sndevent.c). Before parsing the cues, it calls ScanDefineNumeration to build a symbol table of all the DEFINE tokens throughout the script. This is so we can find the appropriate number from the string tokens in the cues themselves. Next, it reads through the script line by line, matching the command keywords discussed above with a set of defined Soundtrack Manager tokens. Outside of comments and PRINT statements, each command begins with the CUE keyword. The CUE token and number are stored in the RAM image, and a placeholder for the offset to the next CUE is written. All CUEs are initially enabled.

Following the CUE keyword and number are the commands of the cue itself. Using a large switch statement, SNDEVENT parses each command and its arguments. Any trailing optional arguments are first set to their default values and then overridden if specified by the audio artist. At the end of every command, the current binary running position of the output file is set to a long word (4-byte) boundary to accommodate those platforms that require such alignment for reading and seeking purposes. When there are no more commands, the file is written out and closed.

There can be any number of binary script files used in a game. That decision is left totally up to the audio artist creating the content and scripts. The game programmer has to know what the file name of the binary configuration file is, when to load it and what cue numbers to call when. Thereafter, the script files and content can change, and the game programmer is none the wiser.

Now we're talkin'

As with many other features, the Soundtrack Manager must be told to support scripts. Defining `__SLIB_MODULE_SCRIPT__` in the `modules.h` header file does this. There are three functions in the public programming interface for scripts within the Soundtrack Manager (all in `\Code\Sound\Script.h`). The first is `SMLoadBinaryConfig`. This takes a single argument, the file name of the binary script file to load. The Soundtrack Manager can only have one script loaded and active at any one time. Therefore, this command automatically unloads any previous binary script file, and loads the specified file into the static module global `ConfigImage`.

The second function in the public script interface is `SMEvent`. This is the main interface function used by the game to issue commands to the Soundtrack Manager. It takes five arguments: the event or `CUE` number, the distance, balance and fader values for the command, and the time at which to play this event. The first parameter is the cue number called by the game (after discussions with the audio team, of course). The next three parameters are used by the application to position the sounds of the cue in the two-dimensional horizontal plane (see above and Chapter 11). The final `msDelay` parameter is used to set up a delay event. This is simply a way to schedule an event to happen some number of milliseconds in the future.

The last function in the public script interface is `SMReset`. This functions exactly as the `RESET` command above. All sounds loaded and playing under script control are stopped and unloaded.

Closing the loop

The final piece of the Soundtrack Manger's scripting functionality is the code within the sound engine itself that reads the binary script file and finds and executes the desired commands. This magic happens in the private `SMPlaySoundEvent` routine, called by the public `SMEvent` function. Within this routine, we search the loaded `ConfigImage` for the event number to play. If it's not in the binary script file, it simply returns `FALSE`. If the cue number is found, however, that's when things get interesting. If the `CUE` is enabled, we enter a large switch statement wherein each of the script commands is mapped to a corresponding script handling function using `case` statements. The command arguments are read from the binary script file, and the appropriate handler function is called. These script handlers are nothing more than wrappers to the native public Soundtrack Manager API.

As a quick example, let's take a look at what happens for the `FADEMUSIC` command. It gets parsed within `SMPlaySoundEvent` as follows:

```
case TFADEMUSIC:
  volume = *binRunPos++;
  pan = (INT8)*binRunPos++;
  numSecs = *binRunPos++;
```

```
FFadeMusic(volume, pan, numSecs);
binRunPos =
  (UINT8 *) ((UINT32) (binRunPos+3) & 0xfffffffc);
break;
```

This command takes three arguments: the target volume and pan position of the fade, and its duration in milliseconds. `binRunPos` is the running binary position within the currently loaded script file, and all script command information is retrieved through this pointer. After getting the command parameters, the script handler `FFadeMusic` is called. As you can see below, `FFadeMusic` is simply a wrapper function to public `SMSetMasterVolumeFade` (in `\Code\Sound\Volume.c`):

```
BOOLEAN FFadeMusic(UINT16 targetVolume, INT16 targetPan,
                   UINT8 numSecs)
{
    if (SMSetMasterVolumeFade (MusicMasterVolumePtr,
          targetVolume, targetPan, numSecs*1000)) {
       return(TRUE);
    }
    return(FALSE);
}
```

Finally, `binRunPos` is updated to the next longword boundary, and we're ready to process more commands.

As simple as 1-2-3

If no binary script file has been loaded by the game or application, the Soundtrack Manager attempts to load a file called `music.bin` the first time it receives an `SMEvent` call. As mentioned above, if there is a binary script file already loaded when the call comes in to `SMLoadBinaryConfig`, the existing script is unloaded in favor of the new one. All this is to say that the programmer need only ever call `SMLoadBinaryConfig` and `SMEvent` to use scripts in a game.

Summary

This chapter presented a cue-based text interface that provides direct control of all the Soundtrack Manager's services. This fulfills our last major objective: that the full power of the sound engine be available to an audio artist to create the game's soundtrack. All the script command syntax and their parameters were presented, as well as the scipt compiler utility `SNDEVENT`. Finally, the mechanisms within the audio engine for parsing and responding to `CUE` events were presented.

11 Case Study 3: Audio/visual coordination – I'll be around …

Auditory cues give important information about the environment in a video game. For example, you may not see the monster, but hearing where it is gives you the advantage. It is also true that coordinated visual and auditory displays make a game more compelling than either single sense alone. This chapter describes one communication channel between the game and audio engines providing real-time audiovisual synchronization.

Tommy, can you hear me?

We discern a lot of information about the world around us just by listening. We can tell what kind of space we're in by the character of its reverberation: a commuter train car sounds different than a living room which sounds different from a concert hall which sounds different than being in a cave or out-of-doors. We can judge the location, distance and movement of things by attending to such cues as interaural time and intensity differences, direct to reverberant sound ratios and Doppler shift. In music, characters are often depicted using individual themes or motifs which we can identify as the piece is played. Many a composer, both classical and popular, has made use of this dramatic device. Think of "Peter and the Wolf" by Prokofiev, the Ring cycle by Wagner, the zaniness of the Warner Brothers cartoons scored by Carl Stalling or the Star Wars movie soundtracks of John Williams. Music also has the uncanny ability to transcend time and space, transporting us to far off times and places long forgotten or perhaps only imagined. Where do you go when you hear the theme song to that game or TV show from your childhood? Do you smell the basement of your old house, or remember wrestling your siblings for the controller?

The point is hearing is a powerful, evocative sense that helps to immerse a player in the game's environment. Attaching sounds to characters and objects lets the player know who or what is there just by listening. There is also great synergy in coordinated visual and audio presentations. Dynamically changing the aural landscape in concert with the visual field draws a player further into the game than either one of the senses can do alone.

The Soundtrack Manager uses what it terms "proximity sounds" to provide all the sonic behaviors and actions listed above. Based on the scripting facility of the last chapter, proximity sounds provide a way to dynamically change the volume and apparent lateral position of any previously loaded sound effect, MIDI track, key-mapped sample or CD audio track in the game. The audio artist determines what type of content best suits the character or object, and creates a proximity sound CUE containing one of these four commands. For example:

```
//Load Harley proximity sound
DEFINE SFXHARLEY 10
DEFINE LOADHARLEYPS 99
DEFINE HARLEYPS 100
CUE LOADHARLEYPS
  LOADSFX SFXHARLEY audio\bikes\harley1.wav
  MAPSFX SFXHARLEY 4 50 70 60 127 0 1
ENDCUE
//Harley proximity sound cue. There should be nothing
// else but the proximity sound command in this cue.
CUE HARLEYPS
  PLAYMIDISFX 4 60 100
ENDCUE
```

As with all other audio cues, the audio artist must tell the game programmer what cue number to use for this proximity sound. But the similarity ends there. In the previous chapter, you read how the game programmer calls SMEvent to trigger a sound cue. From that moment forward, all responsibility for the sounds in that cue rests entirely within the Soundtrack Manager. In the case of proximity sound cues, however, the responsibility for the sound rests equally with the programming and audio teams. The audio engine retains no information about game objects or their positions, but knows how to change the volume and position of sounds. The game engine, on the other hand, has no idea how to move sounds around, but does know of characters and objects and their relative positions to the player within the game. These two information streams come together in proximity sounds. The game tells the Soundtrack Manager where the object is, and the Soundtrack Manager renders the sound in that location.

Proximity sound support is enabled by defining __SLIB_MODULE_PROXIMITY-SOUND__ in modules.h and recompiling. This exposes three additional proximity sound functions in the Soundtrack Manager's public interface: SMNewProximity-Sound, SMProximity and SMDisposeProximitySound. The game engine associates a sound with some visual (or unseen) entity by calling SMNewProximity Sound with the appropriate cue number. The Soundtrack Manager returns a PROXIMITYSOUND pointer to the game engine. Using this pointer, the game conveys to the sound engine the object's relative position by calling `SMProximity`, and the Soundtrack Manager plays the sound in the requested location. When this visual and aural association is no longer needed, the game ends the relationship by calling `SMDisposeProximitySound`.

Speaking of PROXIMITYSOUNDS ...

`SMNewProximitySound` takes a single argument: the artist-supplied proximity sound cue number. It searches for the specified CUE in the currently loaded binary script file, and checks for one of the four legal script commands (listed above). It creates a new PROXIMITYSOUND object for each type of sound source, which looks like this:

```
typedef struct _proximitysound
{
  UINT32          theEvent;
  UINT16          command;
  UINT32          framesToLive;
  FIXED20_12      volumeScalar;
  union {
    PSMIDISFX       theMIDISFX;
    PSSFX           theSFX;
    PSMIDITRACK     theMIDITrack;
    PSCDTRACK       theCDTrack;
  } psu;
  UINT8           distance;
  INT8            balance;
  INT8            fader;
  struct _proximitysound *prev;
  struct _proximitysound *next;
  void *platformData;
} PROXIMITYSOUND;
```

The first two fields identify the cue number and command type used to make this proximity sound. Again, the four legal proximity sound sources are: a sound effect, a MIDI track, a key-mapped sample or a CD audio track. These correspond to the PLAYSFX, UNMUTEMIDITRACK, PLAYMIDISFX and PLAYCDAUDIO script commands, respectively. (The entire list of internal audio command types can be seen in the global `ScriptKeywordConstants` enum in `\Code\Sound\Script.h`.) The Soundtrack Manager uses this command type to decide how to treat a given PROXIMITYSOUND in subsequent functions.

The next member of the PROXIMITYSOUND structure is `framesToLive`. This is a counter that tells the Soundtrack Manager whether this PROXIMITYSOUND is active or not. Each time the game updates a proximity sound, `framesToLive` gets reset to

MAXFRAMESTOLIVE (set to 32 in `\Code\Sound\Proxsnd.h`). framesToLive gets decremented once every timer interrupt in UpdateProximitySounds (one of the list of functions in UpdateProcList, discussed in Chapter 3). If the game does not update the proximity sound again by the time framesToLive reaches 0, the Soundtrack Manger will stop or mute the sound as necessary for the given sound type. At 8 ms per frame, this means the game must update the sound within 256 ms or the sound will no longer be heard. This translates into an update rate just under 4 Hz.

volumeScalar is a 32-bit fixed-point quantity that functions as a volume limiter. Each time a PROXIMITYSOUND is created, volumeScalar is initialized to 1 with 12-bits of fractional resolution: $1 \ll 12$ or 0x1000 (in InitializeProximitySound in `\Code\Sound\Proxsnd.c`). However, the maximum volume of the proximity sound is set in the CUE itself using the volume parameter of the individual script command. In the example above, the maximum volume of the WAV file is set to 100 by the PLAYMIDISFX command in the HARLEYPS cue. Therefore, as the game updates the position of this PROXIMITYSOUND, it will never get any louder than 100, no matter how close to the player it gets. This is a handy way for the audio artist to tweak the balance of the sounds in a game without having to re-engineer the source material itself. A few changes to the script file, and it's done.

Next in PROXIMITYSOUND is a union of the four supported script command data structures. Each time the Soundtrack Manager touches a PROXIMITYSOUND, it does so in a content-specific manner. That is, the information it needs to play a key-mapped MIDI sound effect is different from the information it needs to control the playback of a MIDI track. The Soundtrack Manager uses the command data member to access the appropriate structure in the psu union. The four data structures themselves are called PSMIDISFX, PSSFX, PSMIDITRACK and PSCDTRACK, and are defined in `\Code\Sound\Proxsnd.h` for the code-curious among you.

The next three members are distance, balance and fader. The values of these quantities come from the game engine (in the SMProximity call) and represent the current location of the sounding object relative to the listener. distance ranges from 0 (close) to 255 (far away). balance serves as a pan controller, taking on values from 127 (far right) to −128 (far left), and fader indicates the forward and backward position of the sound, ranging from −128 (behind) to 127 (in front). The Soundtrack Manager adjusts the volume and horizontal position of the sound using these three parameters, essentially placing the sound anywhere in the horizontal plane.

The prev and next data members are used to manage a global linked list of allocated PROXIMITYSOUNDs, called ProximitySounds. These variables allow us to dynamically allocate, use and free proximity sounds as they are created, updated or disposed of by the game.

The final member of the PROXIMITYSOUND structure is platformData. As its name implies, this is used to store any platform-specific information for each proximity sound. Each piece of DA used in scripts is encapsulated inside an SFX data structure. These are stored in a static array of SFX pointers, called SFXFiles. Each time a PROXIMITYSOUND is created using the PLAYSFX or PLAYMIDISFX command, a pointer to the associated digital audio `SFX` is stored in `platformData`. When it is time to update the position of a sound, having this pointer at hand is much more efficient than searching through the entire `SFXFiles` array. It provides a way to quickly and easily update the sound when we have to play, stop, adjust the volume or pitch bend of the underlying DA samples.

You move me

Upon successful creation, `SMNewProximitySound` returns a new `PROXIMITYSOUND` pointer to the game. The sound is initially silent, awaiting more instructions from the game. Those instructions come in the form of calls to `SMProximity`. This function takes five arguments: a `PROXIMITYSOUND` pointer, `distance`, `balance` and `fader` values specifying the position of the sounding object, and a `pitch` parameter. The `PROXIMITYSOUND` pointer is essentially a handle to the sound. The `balance` parameter is used to set the pan position of the sound, while `distance` is used to calculate the overall volume of the sound. This is scaled by the `volumeScalar` parameter of the `PROXIMITYSOUND` structure (discussed above). `fader` describes the front/back position of the sound. For each of the four types of sounds, the static `VolumeDistortionTable` (in `\Code\Sound\Proxsnd.c`) is used to further reduce the volume of the sound when it is located behind the listener (i.e. when `fader` is <0). This is done to mimic the psychoacoustical effect of sounds being softer when they are located behind as opposed to in front of a listener at the same distance.

The final argument to `SMProximity` is `pitch`. This parameter is used to simulate Doppler shift on sound effects and key-mapped MIDI sound effects. An appropriate amount of pitch shifting is applied to those sounds using the difference between the last and current frame's `distance` parameter. The amount of pitch shift in the Soundtrack Manager has been tuned totally by listening trials. That is, it is meant to be an interesting sonic addition, not a scientific simulation. While Doppler shift applies to all sounds in the real world, the sound engine's pitch-shift effect is not applied to CD audio or MIDI tracks. This is because music tends to sound rather bizarre and becomes distracting when it is pitch-shifted. Therefore, this is one case where simulating real-world behavior detracts from the overall game experience.

As mentioned above, the game has to call `SMProximity` within 256 ms for each `PROXIMITYSOUND` or the Soundtrack Manager will mute or stop the sound. This is done so that the game only has to update those sounds it thinks are within earshot. It does not have to update sounds that are very far away and hence inaudible. When a sounding object gets far away, the game simply stops calling `SMProximity`. Should the player ever get close to it again, or it to her, the game resumes calling `SMProximity` with updated parameters, and the Soundtrack Manager responds appropriately. This reduces unnecessary processing overhead and presents the player with only the most important sonic information.

Proximity sounds all behave a little differently from one another, depending on the underlying audio source. Those involving MIDI sound effects or key-mapped MIDI samples are not considered to be music. Therefore, they are started on the first call to SMProximity and stopped when framesToLive reaches 0. Subsequent calls to SMProximity restarts those sounds, and the cycle continues until the game calls SMDisposeProximitySound for that resource. Proximity sounds using a CD audio track, on the other hand, are treated as music. In this case, we don't want to hear the same section of the music over and over again. Instead, we want to play the track through maintaining the integrity of the music itself. The selected track is started the first time the game calls SMProximity, and it keeps looping until the game calls SMDisposeProximitySound to stop it.

The capabilities of the NMD MIDI file format come into play for the UNMUTEMIDITRACK command. Also considered music, proximity sound MIDI tracks are not stopped when the game stops calling SMProximity. They are muted, but still keep track of the elapsed time while silent. When the object gets close enough to be heard again, the game calls SMProximity with the new position. The MIDI track is unmuted, and resumes playing at the volume and pan of the new orientation. Not only are all subsequent MIDI notes played, all notes turned on while the track was muted but also those which have not yet been turned off are restarted due to the magic of CATCHUPEVENTs (see Chapter 6). Standard MIDI files do not support this behavior.

The MIDI file containing proximity sound tracks must already be playing and looping when the game calls SMProximity. The Soundtrack Manager will not start up the MIDI file if it is not playing at that moment. Therefore, the audio artist must create one or more cues to load and play the MIDI file, and immediately mute any PROXIMITY-SOUND tracks. As with key-mapped sound effects, the maximum volume of the MIDI track proximity sound is set by the volume parameter in the LOADMIDI command itself. Therefore, as the game updates the position of a MIDI track PROXIMITY-SOUND, it will never get any louder than what was specified in the original load cue, no matter how close the player comes to that object.

We got the beat

The Soundtrack Manager is also able to communicate beat information to the game engine. This is a very useful technique that can be used to synchronize animations to the beat of the music. In this case, the music drives the game, not the other way around. This could be done in concert with MIDI track proximity sounds, but does not have to be. Any MIDI file will suffice.

When a MIDI file is played, either through the Soundtrack Manager API or, preferably, via the PLAYMIDI script command, the sound engine initializes two global variables, TicksPerFrame and BeatAccumulator (in \Code\Sound\MIDIFile.c). Each MIDI file contains at least one meta-event tempo message specifying the number of microseconds per quarter note in the music. From that quantity is determined the number of quarter notes per second. Each quarter note, or beat, of the MIDI file is further subdivided into 4096 ticks per beat. For the purposes of synchronization, the number of ticks per timer interrupt, or frame, at the given tempo is calculated by multiplying the number of quarter notes per second by the number of ticks per beat and dividing by the Soundtrack Manager frame rate (see below). This the value stored in `TicksPerFrame`. The value of this quantity will change on subsequent meta-event tempo messages, but will always reflect the current tempo of the music:

```
FIXED20_12 qps, tpf;
qps = (FIXED20_12)((1000000 * 512 /
  masterPlayer->MIDIFile->uSecsPerQuarterNote) * 8);
tpf = (FIXED20_12)(TICKSPERBEAT * qps /
  (SMGetFrameRate()>>12));
TicksPerFrame = tpf;
BeatAccumulator = 0;
```

`BeatAccumulator` keeps a running total of the number of ticks as the MIDI file plays, and is initialized to 0. `UpdateMIDIFiles` increments `BeatAccumulator` by `TicksPerFrame` on every timer interrupt.

Both `TicksPerFrame` and `BeatAccumulator` are 32-bit fixed-point integers. This is done for several reasons. First, it retains the accuracy of a fractional number of quarter notes per second and ticks per frame. Ignoring those fractional pieces would introduce quantization error that would eventually cause our tick count to be out of sync with the music. Fixed-point quantities are also used because some platforms do not support floating-point operations. Even on those platforms that can deal with floating-point numbers, integer operations are generally faster. The accuracy of these fixed-point integers is only needed internally, however. When it comes time to report the number of ticks to the outside world, the Soundtrack Manager converts `BeatAccumulator` into a normal 16-bit integer, and places the result in `BeatModulo` (in `\Code\Sound\MIDIFile.c`):

```
void UpdateMIDIFiles(void)
{
  UpdateNoteTable();
  UpdateMasterPlayers();

  //Increment beat counter
  BeatAccumulator += TicksPerFrame;

  //Convert and update tick count
  BeatModulo =(UINT16)(BeatAccumulator >> 12);
}
```

`BeatModulo` is an exported global variable that can be used by the game to coordinate the visual display with the music. For instance, using the fact that there are 4096 ticks per beat, the application can perform a simple modulo operation to check for the next beat. Or it can divide `BeatModulo` by `TICKSPERBEAT` to determine the number of beats so far. It all depends on what the application wants to do with this information.

A sample case will help to make this beat subject clear. Assume a meta-event message that sets the tempo to be 500 000 microseconds per quarter note (the beat). Dividing the number of microseconds in a second by this quantity yields (2) beats per second: 1 000 000 / 500 000 = 2. The number of ticks per second is found by multiplying beats per second times `TICKSPERBEAT`: 2 beats per second * 4096 ticks per beat = 8192 ticks per second. This is finally converted into ticks per frame by dividing that quantity by the number of frames per second of the Soundtrack Manager. Given an 8 ms timer interrupt, this yields a frame rate of 125 Hz. Therefore: 8192 ticks per second / 125 frames per second = 65.536 ticks per frame. Since we prefer not to use floating-point numbers, `HandleMIDIFileEventSetTempo` shifts all of these numbers left by 12 bits, transforming them into fixed-point quantities. This is effectively the same as multiplying our result by 4096 and truncating: 65.536 * 4096 = 268 435.456 which yields 268 435 ticks per frame in fixed-point representation. The actual left-shifting by 12 is done in stages in `HandleMIDIFileEventSetTempo` to retain as much accuracy as possible in the calculation:

```
FIXED20_12 qps, tpf;
qps = (FIXED20_12)((1000000 * 512 / uSpq) * 8);
tpf = (FIXED20_12)(TICKSPERBEAT * qps / (frameRate>>12));
if(tpf != TicksPerFrame) TicksPerFrame = tpf;
```

In the above code, `frameRate` is already a `FIXED20_12` quantity and must be right-shifted 12 bits before the final ticks per frame value is calculated. If this result is not equal to the current `TicksPerFrame` value, that quantity is reset to the new tempo. Once again, `UpdateMIDIFiles` increments `BeatAccumulator` by `TicksPerFrame` on every timer interrupt. Finally, the Soundtrack Manager converts `BeatAccumulator` into a 16-bit integer and places the result in `BeatModulo` for the game.

Summary

In this chapter, we presented an important and powerful technique for coordinating the audio and visual portions of a game called "proximity sounds." We showed how the game programmer and audio artist could collaborate to attach sounds to arbitrary game objects using a small number of script commands. We further demonstrated how the music could drive the game by providing the current beat information to the application.

12 Future topics and directions – where do we go from here?

The focus throughout this book has been to design and build an interactive audio system for games. The Soundtrack Manager presents one such system, and it would be nice if that were the entire, definitive story. Yet the fact of the matter is that while the system presented herein is a good start, there is a great deal more work that has to be and is being done. In this chapter, we explore some of those remaining tasks and subjects.

Interactive audio everywhere

The sound engine described in this book primarily discusses desktop personal computers and game consoles. However, following on our major goals stated way back in Chapter 3, there's every reason to make interactive audio operations available on other platforms. So long as some basic low-level audio services exist, the Soundtrack Manager can potentially run anywhere. High-level musical behavior remains the same regardless of the platform, from desktop machines to handhelds and cell phones. A powerful, flexible and reconfigurable audio engine can make any game on any platform more fun. To run on smaller machines, it is clear that the Soundtrack Manager would have to be optimized to reduce its code size. Some of its more sophisticated components may consume too many resources on smaller machines, so it would have to be reconfigured appropriately. But portable and wireless devices are becoming more powerful all the time. For instance, many cellular phones already contain embedded software wavetable synthesizers. The additional processing overhead of an interactive audio system is minimal by comparison to such processes, and adds significant value to the overall game experience.

For those games targeted for release on multiple platforms, cross-platform operability presents many challenges. Audio file formats and native sound capabilities vary widely across machines, which can make porting from one platform to another a big headache. Traditionally, the game engine itself had to be modified to call the platform-dependent audio API that may or may not have all the support the game requires. Such low-level dependencies only serve to detract from the game's aural experience, and introduce inconsistencies among its different versions. The Soundtrack

Manager goes a long way toward solving these problems by isolating platform-specific operations from abstracted behavior. It has been shown that by addressing a platform's underlying audio API through an individual native adaptor layer, different platforms and capabilities can be accommodated without sacrificing desired behavior. Writing this adaptor layer may seem like a lot of work at first, but this one-time programming effort yields interactive audio means across platforms for all current and subsequent products.

While the Soundtrack Manager is a reasonably sophisticated audio system, there are some game audio processes it does not address. For example, other real-time sound synthesis schemes, either in the form of physical modeling algorithms or those within the MPEG-4 standard, are not accommodated. There is also no explicit use of spatialized sound or positional three-dimensional audio, although hooks are in place within the SMEvent and SMProximity calls. While the Soundtrack Manager implements some basic sound effects in the FPMIX version of the library, it does not have any generalized way of addressing native sound effects or their parametric controls across platforms. This last subject is not just a game audio problem. Many manufacturers have products to do effects processing, but there is no standardized way to address and manage the algorithmic control parameters of these effects.

Another one of the Soundtrack Manager's major design goals was to put high-level musical behaviors into the hands of an audio artist. This speaks of the tremendous need for a new, cross-platform audio authoring tool. It should be constructed to facilitate how an artist thinks about and conceptualizes an interactive audio soundtrack. Processes for interactive mixing and playback scheduling should be developed to facilitate a game's need to dynamically order and present the audio content. The authoring tool should be integrated with a real-time audio rendering engine, akin to the Soundtrack Manager, running directly on the game's target platform along with the game itself. Game development tools could even go so far as to simulate or virtualize the behavior and sonic performance of a specific game platform on some larger workstation. The point is that the ability to hear how the audio will sound and behave on the target platform in the context of the actual game is invaluable to the quality of the final product. Such a system should also be able to ship all the audio content and along with all its performance and articulation data inside a single standardized, efficient, flexible and non-proprietary media container. Having a common media wrapper supports interoperability, and could greatly smooth the progress of content distribution for all manner of entertainment devices.

For all of the positive things presented in this book, the reader should be mindful that this is only one person's perspective and solution. For the field of interactive audio to grow, all issues and their potential solutions must be considered and discussed by a larger group. Interactive audio is an idea whose time has come, but there is still a need for a common vocabulary and standardized functionality. I am pleased to report that there is just such an effort underway in the Interactive Audio Special Interest Group (IA-SIG; www.iasig.org). Within this organization, the Interactive XMF Working Group is charged with defining both the capabilities and supporting data structures of a wide-ranging and highly capable interactive audio engine. In addition, all the data necessary for the performance and real-time control of the interactive soundtrack is to be wrapped inside the open eXtensible Music Format (XMF), available from the MMA.

Let's get together

Beyond the need for systems to create and play back interactive audio soundtracks, there are other areas of game production and reproduction that require attention. For example, though games have blossomed into a $10 billion annual industry, there are no industry-wide recommended practices for either the production or reproduction of game audio. Movie companies pay a lot of attention to the audio quality of their products. Film soundtracks are mixed in highly specified and controlled environments, and the audio is re-mixed for each release format (LaserDisc, video tape, DVD, etc.). Game houses, for the most part, do not follow this model. The time and resources given to audio in games do not allow for this kind of attention and reflect the general undervaluing of the power of sound and its contribution to the overall success of a game. That said, it is up to the audio practitioners in the field to keep the pressure on and push for positive change. By defining, developing and adhering to accepted audio engineering and mixing practices, we will demonstrate, improve and ensure the quality of all game soundtracks.

There are many questions we need to answer to transform and grow game audio into the high-quality and compelling experience it can be. What should the parameters of a calibrated audio production environment be? How should it behave acoustically? What are the preferred performance characteristics of the individual system components? What are the recommended ways to produce a multichannel game audio soundtrack, and how is that different from movies? What are the audio performance characteristics of the various game consoles, and what should be done on the production side to make the audio sound the best that it can on each of those platforms? What are the best working methodologies for making game audio?

Games @ home

Moving past production and into the home, there are similarly no recommended practices or guidelines about how to set up a listening room for games or what kind of performance to expect from a home arcade system. Simply because we can hook up game machines to existing home theater systems doesn't mean we should. The audio reproduction requirements of a multichannel, multiplayer racing or fighting game are different from those of passive movie-watching. While both movies and games attempt to place you in some virtual auditory environment, games are much more visceral and require different performance. Movies tend to immerse the listener in a diffuse and decorrelated sound field, but in a game it is important to be able to hear precisely where you and your enemies or competitors are at any given moment. Your survival and success depend on it.

Home theater systems are engineered and installed such that there is one best listening position, or "sweet spot," accommodating only one or two people. Everyone else outside this ideal listening position experiences an impoverished audio presentation. This is problematic for games when all players share one system with a single screen. Only one or two players get the full audio effect of the game. Home theater systems also completely ignore the vertical dimension, reproducing sounds primarily on the horizontal plane. There is no use case for people changing positions when they watch a film. However, game players are not so sedentary and often stand up or move around while playing the game. When they do this, they are no longer in the prescribed auditory field of the system and their aural environment instantly, and negatively, changes.

Many games now utilize some form of three-dimensional sound processing. The aim of this is twofold: to envelop the player in a virtual auditory environment and to deliver precise localization cues for various sounding objects. There are many well-known problems with these kinds of simulations, depending on the mode of presentation. Over headphones, three-dimensional sound processing can move sounds outside the head, giving an increased sense of spaciousness. Also, the image does not change as the player moves around as the headphones move with her. Frontal imagery is not very good, however, and sounds tend to fall toward the rear. Over conventional stereo loudspeakers, sound immersion is reasonably successful, and it is possible to achieve some vertical effects. Frontal imagery is well defined, but it is next to impossible to place sounds behind the listener. Sounds also collapse into a single speaker if you move off the centerline of the speakers, due to the precedence effect. Multichannel systems provide very good envelopment, and are not as susceptible to the precedence effect as two-channel speakers. Lateralization works pretty well, especially in discrete multichannel systems where each channel gets its own full-range speaker. It is easy to have sounds appear at the sides or behind the listener, as there are physical transducers there. Here again, these effects are best for the listener in the sweet spot, and all three-dimensional localization cues are collapsed onto the horizontal plane.

With what shall we fix it

One solution is to provide a dedicated multichannel audio playback environment for each player. Depending on how this is implemented, this could put an additional processing burden on the game to uniquely render the audio for each player based on his or her character, location and perspective. This would be a difficult sell to a producer if it meant any degradation of the video performance, and it would certainly introduce extra cost to the consumer for multiple multichannel setups. The social, acoustic, ergonomic and networking challenges for such a setup are great, but can be met with today's technology. But while this solves the problem of squeezing everyone into a tight cluster in front of a single screen, it still doesn't address the horizontal plane problem.

Another new and promising development in multichannel sound reproduction is a technology known as wave-field synthesis. Here, tens of loudspeakers are arrayed about the listening environment to simulate the sound field of one room inside another. Work in this area has initially been concentrated on the study and reproduction of sound fields from existing rooms and halls. The technique has recently been extended to include synthesis of arbitrary and nonphysical spaces, which could be ideal to the needs of games. Precise localization of individual sound sources is still a bit lacking in these simulations, but the precedence effect is almost completely defeated, meaning you can be anywhere in the room, facing in any direction and still experience a high quality, enveloping sound. The demonstrations and possibilities of this technology are striking.

From the descriptions above, it is clear that there is no completely satisfying solution when it comes to presenting three-dimensional auditory illusions. I believe a new paradigm for totally immersive audio experiences is required. The most successful alternative would be to broadcast a diffuse, enveloping soundstage using multichannel speakers, and have each person don a set of hybrid assistive listening devices to present individual synthesized localization cues. These devices would take the form of earphone transducers sitting out from the head and firing into the concha (the depression leading into one ear canal). Studies have shown that three-dimensional sound illusions are more convincing when a person's own head-related transfer functions (HRTFs) are used. In the system described here, players would use their natural localization hardware, their own ears, to interpret the synthesized three-dimensional sound cues delivered by the game while simultaneously being immersed in a room full of sound. Such listening devices exist only in research labs at the moment, and are not necessarily being targeted for games. Here is a tremendous opportunity for some discerning and willing manufacturer.

What all this adds up to is that for all the money and attention given to games, there is no reliable way to ensure that the end user is hearing what the game audio artist intended. I believe we must define standards and recommended practices across all facets of the game industry. Only through this kind of attention and adoption will game audio be consistent and of high quality.

The Audio Engineering Society (www.aes.org) has also formed a Technical Committee on Audio for Games to address just these kinds of technical issues to move game audio forward. Again, it will require the participation of the community of game audio practitioners and manufacturers of all stripes to be a success. But I am confident it will happen.

The king is dead – long live the king

Historically, game development projects have been driven by a few, self-selecting individuals who had a good idea and went out, or into someone's basement, and built it. However, the model wherein a few individuals do everything that needs to be done to make a particular game is fast becoming a thing of the past. Games are becoming more and more complex, and the effort, skills and resources necessary to develop and ship a title simultaneously on multiple platforms worldwide are beyond the scope of most small houses. Therefore, the game industry has consolidated to the point where fewer and fewer large companies dominate the game production and publication marketplace, consuming and acquiring all in their path.

I believe it is the very success of games that will cause a paradigm shift in this current monolithic landscape, for two reasons. First, sophisticated third-party development tools are emerging as companies make it their core business to provide the latest and greatest technology across many different areas of game development. Such programs that once dominated were only developed in-house. But as games have advanced, the resources and costs associated with their continued maintenance and development have become prohibitive. If a company can buy the tools it needs to do production off-the-shelf, which can do more than their in-house tools for less money, there's no question which way they'll go. Second, it used to be that the only place one could learn about making games was to land a job in a game house, and pay attention. This is no longer the case. More and more educational institutions are offering courses in all facets of game development and design. They offer students the training they need on these third-party development tools, and are providing a ready-made workforce for games that we have never seen before.

Consequently, I believe the existing model where people work for a specific game development house is ultimately doomed. I see the industry moving more toward the model of films, where teams of individuals, each highly specialized, come together for a particular project and disband when that project is over. I see a culture and workforce of independent production companies and individuals, each contracting their services with the studios on an as-needed basis. The days of everyone waiting on the lead programmer for his or her slice of attention are numbered, too. As game machines become more powerful and the use of third-party integrated development tools takes off, game development will no longer be primarily about the programming. Games will always make use of and exploit the latest technology, but the focus will shift away from the core technology of an individual platform to the broader use of technology across platforms, to the story, the content, the license and to the sophistication of the production.

Unfortunately, and not atypically, the tools for creating an interactive audio soundtrack for games trail the pack in this futuristic landscape. As mentioned above, there is a tremendous need (and opportunity!) for more audio engines and tools in the industry. Audio programmers who currently toil for a specific game company should be encouraged to write game audio engines to do just that. Companies writing game production suites would do well to include advanced support for audio. Including interactive audio in their integrated game production packages would allow a game developer to create, build and render all content for a game, no matter what the platform. The bottom line is games are heading away from proprietary, in-house solutions and more toward better and better third-party solutions.

Dream Jeannie

As Helmholtz said a long time ago, "But certainly it is more difficult to make a proper use of absolute freedom, than to advance where external irremovable landmarks limit the width of the path which the artist has to traverse." In the story of game audio, perhaps we have done this backwards. We first imposed unreasonable limits on our creativity by the operation of the machines to which we were inexorably tied. It was difficult to advance there, but for a while we were comfortable with its familiarity. That is no longer the case. Interactive audio is the lifting of those artificial constraints, the realizing of more freedom of expression in our chosen audio art. We are now exploring what it means to "make proper use" of this newfound freedom, and only just beginning to figure it out. I believe interactive audio is another channel of human expression that extends beyond computer games to many other multimedia art forms. Whole worlds of creative expression will arise as artists combine human gesture and performance of all kinds to create new immersive and participatory experiences.

Therefore, it is my sincere hope and wish that the interactive audio tools and technologies developed initially for games will not only make better games, but kindle the creativity of individuals across all the arts. Whatever your interest in game audio, let yourself go! Create what is yours alone to do, with conscious intention, and the world will be changed. Cheers!

Appendix A: Soundtrack Manager script commands

Note: All arguments listed in order; optional arguments in brackets.

//
Comment indictor. Used at beginning of line only. For script readability, ignored in output binary file.

PRINT
Display a status string during script compilation. Ignored in output binary file.
Example: PRINT "Compiling Level 1 cues"

CUE cueNumber
Command keyword indicating the start of a new sound event cue definition.
Arguments:
    cueNumber – number of this CUE; used by the game or application to execute the
        audio commands in this CUE.
Example: CUE 16 or CUE BRANCHX

ENDCUE
Command keyword indicating the end of a current sound event cue definition.

ENABLECUE cueNumber
Enable the specified cue number.
Arguments:
    cueNumber – number of the CUE to enable
Example: ENABLECUE 42

DISABLECUE cueNumber
Disable the specified cue number.
Arguments:
    cueNumber – number of the CUE to disable
Example: DISABLECUE NINETIMESSIX

DELAYCUE cueNumber msDelay [distance] [balance] [fader]
Schedule the specified cue number to be executed some number of milliseconds in the future.

Arguments:
   cueNumber – number of the CUE to delay
   msDelay – millisecond delay before the specified cue number executes
   distance – how far away the sounds are in cueNumber (0 = close, 255 = far);
     default 0
   balance – pan position of the sounds in cueNumber (−128 left, 127 right);
     default 0
   fader – front/back position of the sounds (−128 behind, 127 front); default 0
Example: DELAYCUE 100 1000 128 - 32 64

DEFINE textToken number
Attach a meaningful text symbol to a number.
Arguments:
   textToken – Text string to use in place of the following number
   number – Number represented by textToken
Example: DEFINE MonsterMash 100

RESET
Stop and unload all sound resources and structures loaded by any previous script
commands.

LOADSFX sfxNumber relativePath [volume] [pan] [masterMusicFlag] [streamFlag]
Load a chunk of digital audio into the Soundtrack Manager.
Arguments:
   sfxNumber – number to which the sound is to be assigned
   relativePath – relative path (from `SoundPath`) of the sound file being loaded
   volume – desired volume of the sound (0 silence, 127 max gain); default 127
   pan – desired pan location of the sound (−128 left, 0 center, 127 right);
     default 0
   masterMusicFlag – sound is part of game music track (0 = no, 1 = yes);
     default 0
   streamFlag – sound should be streamed from disk (0 = no, 1 = yes);
     default 0
Example: LOADSFX 0 sfx\growl.wav 127 0 0 1

UNLOADSFX sfxNumber
Unload a chunk of digital audio from the Soundtrack Manager.
Arguments:
   sfxNumber – number of previously loaded sound effect
Example: UNLOADSFX GROWL

PLAYSFX sfxNumber
Play a sound effect.
Arguments:
   sfxNumber – number of previously loaded sound effect
Example: PLAYSFX 0

STOPSFX sfxNumber
Stop a sound effect.
Arguments:
   sfxNumber – number of previously loaded sound effect
Example: STOPSFX 6

LOOPSFX sfxNumber
Loop a sound effect.
Arguments:
   sfxNumber – number of previously loaded sound effect
Example: LOOPSFX BGTRACK

UNLOOPSFX sfxNumber
Unloop a sound effect.
Arguments:
   sfxNumber – number of previously loaded sound effect
Example: UNLOOPSFX BGTRACK

FADESFX sfxNumber [volume] [pan] [duration]
Fade a sound effect to the specified volume and pan location over duration seconds.
Arguments:
   sfxNumber – number of previously loaded sound effect
   volume – desired volume of the sound (0 silence, 127 max gain); default 0
   pan – desired pan location of the sound (−128 left, 0 center, 127 right); default 0
   duration – length of fade in seconds (0 min, 255 max); default 4
Example: FADESFX BG_MUSIC1 127 0 10

SFXVOLUME sfxNumber [volume] [pan]
Set the sound effect to the specified volume and pan location.
Arguments:
   sfxNumber – number of previously loaded sound effect
   volume – desired volume of the sound (0 silence, 127 max gain); default 127
   pan – desired pan location of the sound (−128 left, 0 center, 127 right); default 0
Example: SFXVOLUME BG_MUSIC1 64 - 64

MAPSFX sfxNumber channel lowNote [highNote] [ctrNote] [volume] [pan]
   [monoFlag]
Map a sound effect onto a range of keys on the specified MIDI channel.
Arguments:
   sfxNumber – number of previously loaded sound effect
   channel – MIDI channel onto which the sound will be mapped (0–15)
   lowNote – starting MIDI note number of key range
   highNote – ending MIDI note number of key range; default lowNote
   ctrNote – MIDI note number where sound not transposed; default highNote
   volume – desired volume of the sound (0 silence, 127 max gain); default 127

pan – desired pan location of the sound (−128 left, 0 center, 127 right); default 0

monoFlag – sound is monaural (0 = FALSE, 1 = TRUE); default 0

Example: MAPSFX PIANO1 1 54 64 60 127 0 0

UNMAPSFX sfxNumber channel

Unmap a sound effect from all notes on the specified MIDI channel.

Arguments:

sfxNumber – number of previously loaded sound effect

channel – MIDI channel onto which the sound was previously mapped

Example: UNMAPSFX PIANO1 1

DRUMSFX sfxNumber

Let sound play all the way through, ignoring MIDI note off messages.

Arguments:

sfxNumber – number of previously loaded sound effect

Example: DRUMSFX BASS_DRUM

UNDRUMSFX sfxNumber

Turn off drum behavior for this sound; respond to MIDI note off messages.

Arguments:

sfxNumber – number of previously loaded sound effect

Example: UNDRUMSFX BASS_DRUM

SFXENVELOPE sfxNumber [attackTime] [releaseTime]

Set the attack and release times of a previously loaded sound effect. Sound will be ramped up in volume over attackTime milliseconds in response to a MIDI note on message. Sound will be faded to silence over releaseTime milliseconds in response to a MIDI note off message.

Arguments:

sfxNumber – number of previously loaded sound effect

attackTime – length of attack phase for this sound in milliseconds; default 0

releaseTime – length of release phase for this sound in milliseconds; default 0

Example: SFXENVELOPE PIANO1 10 500

PLAYMIDISFX channel noteNumber [volume] [pan]

Play the sound that was previously loaded onto the specified MIDI channel and note number.

Arguments:

channel – MIDI channel onto which a sound was previously mapped (0–15)

noteNumber – MIDI note number to play; must be in mapped range (0–127)

volume – desired volume of the sound (0 silence, 127 max gain); default 127

pan – desired pan location of the sound (−128 left, 0 center, 127 right); default 0

Example: PLAYMIDISFX 1 60 32 64

STOPMIDISFX channel noteNumber

Stop the stops the sound mapped to the specific channel and key combination.

Arguments:

  channel – MIDI channel onto which a sound was previously mapped (0–15)
  noteNumber – MIDI note number to play; must be in mapped range (0–127)
Example: STOPMIDISFX 9 60

LOADMIDI midiNumber relativePath [volume] [pan] [loopFlag]
Load an NMD MIDI file into the Soundtrack Manager.
Arguments:

  midiNumber – number to which the NMD file is to be assigned
  relativePath – relative path (from SoundPath) of the NMD file being loaded
  volume – desired volume of the MIDI file (0 silence, 127 max gain); default 127
  pan – desired pan location of the MIDI file (−128 left, 0 center, 127 right);
    default 0
  loopFlag – flag indicating desired MIDI file loop status (T = yes, F = no);
    default F
Example: LOADMIDI 0 music\Battle.nmd 100 0 T

UNLOADMIDI midiNumber
Unload a previously loaded MIDI file from the Soundtrack Manager.
Arguments:

  midiNumber – MIDI file number to unload
Example: UNLOADMIDI 0

PLAYMIDI midiNumber
Play a MIDI file.
Arguments:

  midiNumber – loaded MIDI file number
Example: PLAYMIDI INTRO1

STOPMIDI midiNumber
Stop a MIDI file.
Arguments:

  midiNumber – loaded MIDI file number
Example: STOPMIDI INTRO2

LOOPMIDI midiNumber
Loop a MIDI file.
Arguments:

  midiNumber – loaded MIDI file number
Example: LOOPMIDI VAMP

UNLOOPMIDI midiNumber
Unloop a MIDI file.
Arguments:

  midiNumber – loaded MIDI file number
Example: UNLOOPMIDI VAMP

PAUSEMIDI midiNumber
Pause a MIDI file.
Arguments:
   midiNumber – loaded MIDI file number
Example: PAUSEMIDI 100

RESUMEMIDI midiNumber
Resume playing a MIDI file.
Arguments:
   midiNumber – loaded MIDI file number
Example: RESUMEMIDI 8

MUTEMIDI midiNumber
Mute a MIDI file.
Arguments:
   midiNumber – loaded MIDI file number
Example: MUTEMIDI ENGINESOUND

UNMUTEMIDI midiNumber
Unmute a MIDI file.
Arguments:
   midiNumber – loaded MIDI file number
Example: UNMUTEMIDI ENGINESOUND

MIDIVOLUME midiNumber [volume] [pan]
Set the selected MIDI file to the specified volume and pan location.
Arguments:
   midiNumber – number of previously loaded MIDI file
   volume – desired volume of the sound (0 silence, 127 max gain); default 127
   pan – desired pan location of the sound (−128 left, 0 center, 127 right); default 0
Example: MIDIVOLUME SWORD_MUSIC 32 0

FADEMIDI midiNumber [volume] [pan] [duration]
Fade a MIDI file to the specified volume and pan location over duration seconds.
Arguments:
   midiNumber – number of previously loaded MIDI file
   volume – desired volume of the MIDI file (0 silence, 127 max gain); default 0
   pan – desired pan location of the MIDI file (−128 left, 0 center, 127 right); default 0
   duration – length of fade in seconds (0 min, 255 max); default 4
Example: FADEMIDI LEVEL9_OUT 0 0 60

MIDITRACKVOLUME midiNumber trackNumber [volume] [pan]
Set the selected MIDI file track number to the specified volume and pan location.
Arguments:
   midiNumber – number of previously loaded MIDI file
   trackNumber – track number of previously loaded MIDI file

volume – desired volume of the sound (0 silence, 127 max gain); default 127
pan – desired pan location of the sound (−128 left, 0 center, 127 right); default 0
Example: MIDITRACKVOLUME 12 5 64 64

FADEMIDITRACK midiNumber trackNumber [volume] [pan] [duration]
Fade a MIDI file track to the specified volume and pan location over duration
   seconds.
Arguments:
   midiNumber – number of previously loaded MIDI file
   trackNumber – track number of previously loaded MIDI file
   volume – desired volume of the MIDI file (0 silence, 127 max gain); default 0
   pan – desired pan location of the MIDI file (−128 left, 0 center, 127 right); default 0
   duration – length of fade in seconds (0 min, 255 max); default 4
Example: FADEMIDITRACK LEADINMUSIC HAPPY_TRACK 127 0 4

MUTEMIDITRACK midiNumber trackNumber
Mute the specified MIDI file track number.
Arguments:
   midiNumber – loaded MIDI file number
   trackNumber – track number of that MIDI file
Example: MUTEMIDITRACK FIGHT_MUSIC 3

UNMUTEMIDITRACK midiNumber trackNumber [volume] [pan]
Unmute the specified MIDI file track number.
Arguments:
   midiNumber – loaded MIDI file number
   trackNumber – track number of that MIDI file
   volume – desired volume of the MIDI track (0 silence, 127 max gain); default 0
   pan – desired pan location of the MIDI track (−128 left, 0 center, 127 right); default 0
Example: UNMUTEMIDITRACK FIGHT_MUSIC 3 97 32

SEEKTOMIDIMARKER midiNumber markerNum OR markerName
Seek to a given marker within the specified MIDI file. Can specify either a marker
number or marker name.
Arguments:
   midiNumber – loaded MIDI file number
   markerNum – marker number to which to seek
   markerName – marker name to which to seek
Example: SEEKTOMIDIMARKER 10 27
Example: SEEKTOMIDIMARKER CELEBRATION_MUSIC VERSE2

SETMIDIMARKEREVENT midiNumber cueNumber
Select a script command to execute when the specified MIDI file next hits a marker
event.

Arguments:

    midiNumber – loaded MIDI file number

    cueNumber – number of the CUE to execute upon hitting the next marker

Example: SETMIDIMARKEREVENT BATTLE_MUSIC MONSTER_KILLED_CUE

MUSICVOLUME [volume] [pan]

Set the volume and pan location of all music sounds.

Arguments:

    volume – desired volume of the music (0 silence, 127 max gain); default 127

    pan – desired pan location of the music (−128 left, 0 center, 127 right); default 0

Example: MUSICVOLUME 64 0

FADEMUSIC [volume] [pan] [duration]

Fade all music to the specified volume and pan location over duration seconds.

Arguments:

    volume – desired volume of the music (0 silence, 127 max gain); default 0

    pan – desired pan location of the music (−128 left, 0 center, 127 right); default 0

    duration – length of fade in seconds (0 min, 255 max); default 4

Example: FADEMUSIC 0 0 30

APPEND midiNumber listNumber

Append the specified MIDI file onto the desired playlist.

Arguments:

    midiNumber – previously loaded MIDI file number

    listNumber – playlist number onto which to append the MIDI file

Example: APPEND CHASE1 0

REMOVE midiNumber listNumber

Remove the specified MIDI file from the desired playlist.

Arguments:

    midiNumber – previously loaded MIDI file number

    listNumber – playlist number onto which to append the MIDI file

Example: REMOVE CHASE2 2

PLAYLIST listNumber

Start playing the specified playlist.

Arguments:

    listNumber – playlist number for this operation

Example: PLAYLIST 0

STOPLIST listNumber

Stop playing the specified playlist.

Arguments:

    listNumber – playlist number for this operation

Example: STOPLIST SCENE2BG

PAUSELIST listNumber
Pause the specified playlist.
Arguments:
   listNumber – playlist number for this operation
Example: PAUSELIST 4

RESUMELIST listNumber
Resume playing the specified playlist.
Arguments:
   listNumber – playlist number for this operation
Example: RESUMELIST HEAD_CRANK

LOOPLIST listNumber
Loop the specified playlist.
Arguments:
   listNumber – playlist number for this operation
Example: LOOPLIST 10

UNLOOPLIST listNumber
Unloop the specified playlist.
Arguments:
   listNumber – playlist number for this operation
Example: UNLOOPLIST 0

CLEARLIST listNumber
Clear the specified playlist.
Arguments:
   listNumber – playlist number for this operation
Example: CLEARLIST LevelMusic

LISTVOLUME listNumber [volume] [pan]
Arguments:
   listNumber – playlist number for this operation
   volume – desired volume of the playlist (0 silence, 127 max gain); default 0
   pan – desired pan location of the playlist (−128 left, 0 center, 127 right);
     default 0
Example: LISTVOLUME BG_Level10 28 0

FADELIST listNumber [volume] [pan] [duration]
Fade the indicated playlist to the specified volume and pan location over duration
seconds.
Arguments:
   listNumber – playlist number for this operation
   volume – desired volume of the playlist (0 silence, 127 max gain); default 0

pan – desired pan location of the playlist (−128 left, 0 center, 127 right); default 0
duration – length of fade in seconds (0 min, 255 max); default 4
Example: FADELIST MUSICLIST3 127 0 5

SEGUETO midiNumber destList srcList
Command to segue from the currently playing MIDI file in the source playlist to the
specified MIDI file number in destination playlist. The MIDI file number is the number
given to the file with the LOADMIDI command and does not refer to the order of the
files in the destination playlist. List designations can be the same or different.
Arguments:
    midiNumber – loaded MIDI file number
    destList – playlist number to which to segue containing the desired MIDI file
        number
    srcList – playlist from which to segue when the current MIDI file entry is done
Example: SEGUETO FAIRY_MUSIC CASTLE_LIST DUNGEON_LIST

JUMPTO midiNumber listNumber
Start playing the specified MIDI file number in the indicated playlist.
Arguments:
    midiNumber – MIDI file number previously APPENDed to listNumber
    listNumber – playlist number that contains the MIDI file number
Example: JUMPTO 4 2

PLAYCDAUDIO trackNumber
Begin playing the specified CD audio track number.
Arguments:
    trackNumber – track number of the CD to play
Example: PLAYCDAUDIO 5

STOPCDAUDIO
Stop all CD audio playback.

PAUSEGAME
Pause all currently playing MIDI files and build a list of what files to restart.

RESUMEGAME
Resume playback of all suspended MIDI files.

SAVEGAME
Build a special `RELOADGAMECUENUMBER` cue to restore the state of the currently
playing MIDI files.

RELOADGAME
Call the `RELOADGAMECUENUMBER` cue, restarting all MIDI files.

Note: The following two commands are only operational in the FPMIX version of the Soundtrack Manager

SETDELAY [delayVolume] [delayLength] [feedbackStrength]
Set the parameters of a global reverberation effect.
Arguments:
    delayVolume – volume of the delayed sound (0 silence, 127 max gain); default 0
    delayLength – length of delay in milliseconds (0 to 65535); default 0
    feedbackStrength – volume of a recirculated copy of the input sound (0 silence, 127 max gain); default 0
Example: SETDELAY 64 30 32

SETFLANGE [efxVolume] [nSampDelay] [nSampDepth] [sweepFreq] [invertFlag]
Set the parameters of a global flange effect.
Arguments:
    efxVolume – overall volume of the effect (0 to 127); default 0
    nSampDelay – length of effect delay in samples (0 to 1000); default 0
    nSampDepth – depth of the effect in samples (0 to 1000); default 0
    sweepFreq – a non-negative sweep frequency; default 0
    invertFlag – flag to invert recirculated sound (T = yes, F = no); default F
Example: SETFLANGE 100 500 200 5 F

Appendix B: Game audio resources

Included below are some URLs and names of additional audio programming resources and conferences.

Windows DirectX
http://www.microsoft.com/windows/directx/default.aspx

OpenAL
http://developer.creative.com
http://www.lokigames.com

Open Sound System
http://www.opensound.com

Miles Sound System
http://www.radgametools.com/miles.htm

GameCODA – Multi-platform Audio Middleware
http://www.gamecoda.com

FMOD Music and Sound Effects System
http://www.fmod.org

Mac OS X Core Audio
http://developer.apple.com/audio/macosxaudio.html

Beatnik Audio Engine
http://www.beatnik.com

Factor 5/MusyX Sound System
http://www.factor5.com

Game and Audio Developer Conferences:
Audio Engineering Society
http://www.aes.org

MIDI Manufacturers Association
http://www.midi.org

Interactive Audio Special Interest Group
http://www.iasig.org

Game Audio Network Guild
http://www.audiogang.org

Project Bar-B-Q
http://www.projectbarbq.org

Sonify.org
http://www.sonify.org

Game Developers Conference
http://www.gdconf.com

Independent Game Developers Association
http://www.igda.org

Index

 **Focal Press**   **www.focalpress.com**

Join Focal Press online
As a member you will enjoy the following benefits:

- browse our full list of books available
- view sample chapters
- order securely online

Focal eNews
Register for eNews, the regular email service from Focal Press, to receive:

- advance news of our latest publications
- exclusive articles written by our authors
- related event information
- free sample chapters
- information about special offers

Go to www.focalpress.com to register and the eNews bulletin will soon be arriving on your desktop!

If you require any further information about the eNews or www.focalpress.com please contact:

USA
Tricia Geswell
Email: t.geswell@elsevier.com
Tel: +1 781 313 4739

Europe and rest of world
Lucy Lomas-Walker
Email: l.lomas@elsevier.com
Tel: +44 (0) 1865 314438

Catalogue
For information on all Focal Press titles, our full catalogue is available online at www.focalpress.com, alternatively you can contact us for a free printed version:

USA
Email: c.degon@elsevier.com
Tel: +1 781 313 4721

Europe and rest of world
Email: j.blackford@elsevier.com
Tel: +44 (0) 1865 314220

Potential authors
If you have an idea for a book, please get in touch:

USA
editors@focalpress.com

Europe and rest of world
ge.kennedy@elsevier.com